Ḥikāyat Abī al-Qāsim

Edinburgh Studies in Classical Arabic Literature
Series Editors: Wen-chin Ouyang and Julia Bray

This series departs from conventional writing on Classical Arabic Literature. It integrates into its terms of enquiry both cultural and literary theory and the historical contexts and conceptual categories that shaped individual writers or works of literature. Its approach provides a forum for path-breaking research which has yet to exert an impact on the scholarship. The purpose of the series is to open up new vistas on an intellectual and imaginative tradition that has repeatedly contributed to world cultures and has the continued capacity to stimulate new thinking.

Books in the series include:

The Reader in al-Jāḥiẓ
Thomas Hefter

Recognition in the Arabic Narrative Tradition: Discovery, Deliverance and Delusion
Philip F. Kennedy

Counsel for Kings: Wisdom and Politics in Tenth-Century Iran
Volume I: *The* Naṣīḥat al-mulūk *of Pseudo Māwardī: Contexts and Themes*
Louise Marlow

Counsel for Kings: Wisdom and Politics in Tenth-Century Iran
Volume II: *The* Naṣīḥat al-mulūk *of Pseudo Māwardī: Texts, Sources and Authorities*
Louise Marlow

Al-Jāḥiẓ: In Praise of Books
James E. Montgomery

Al-Jāḥiẓ: In Censure of Books
James E. Montgomery

Ḥikāyat Abī al-Qāsim*: A Literary Banquet*
Emily Selove

The Literary Qur'an
Shawkat M. Toorawa

www.euppublishing.com/series/escal

Ḥikāyat Abī al-Qāsim

A Literary Banquet

Emily Selove

EDINBURGH
University Press

Edinburgh University Press is one of the leading university presses in the UK. We publish academic books and journals in our selected subject areas across the humanities and social sciences, combining cutting-edge scholarship with high editorial and production values to produce academic works of lasting importance. For more information visit our website: www.edinburghuniversitypress.com

Edinburgh University Press Ltd
The Tun – Holyrood Road
12 (2f) Jackson's Entry
Edinburgh EH8 8PJ

Typeset in 11/15 Adobe Garamond by
Servis Filmsetting Ltd, Stockport, Cheshire,
printed and bound in the United States of America

A CIP record for this book is available from the British Library

ISBN 978 1 4744 0231 6 (hardback)
ISBN 978 1 4744 0232 3 (webready PDF)
ISBN 978 1 4744 1158 5 (epub)

Contents

Abbreviations

Ḥ Al-Azdī, *Ḥikāyat Abī al-Qāsim al-Baghdādī*, published as Abū Ḥayyān al-Tawḥīdī (attributed), *al-Risālat al-Baghdādiyya*, ed. ʿAbbūd al-Shāljī (Koln: Manshūrāt al-Jamal, 1980).

Q The Qur'an[1]

Notes

1. The Qur'an is an especially difficult work to translate. In this study I usually use my own translations, but where indicated, I use those of some of its most famous translators, according to whichever best serves to illustrate a given point. In references, the first number refers to the chapter, the second number to the verse (e.g. Q 2:1 refers to the first verse of the second chapter of the Qur'an).

Acknowledgements

I would first like to thank Shawkat Toorawa, adviser of my undergraduate thesis at Cornell University, under whose tutelage I became acquainted with the party-crashing character in Arabic literature. I thank Geert Jan van Gelder, whose *Dishes and Discourse* first introduced me to *Ḥikāyat Abī al-Qāsim*, and who is currently, to my extreme pleasure and honour, co-translating and co-editing its manuscript with me. I fondly remember my time reading the *Ḥikāya* together with him and Adam Talib, whom I would also like to thank. I thank Michael Cooperson, adviser of my UCLA Ph.D. thesis, which formed the basis of this book. He gave me both freedom and guidance, both mentorship and friendship; he has continued to assist me even after my graduation from UCLA, and he is my single biggest influence in my articulation and organisation of the ideas presented here. I thank Amy Richlin for many hilarious hours of reading Latin together, including most of the Roman texts mentioned in this study, and also for sharing her experience, her insight, and her dogs with me. I thank Matthew Leigh for commenting on several chapters of this book, and for introducing me to his lovely hound Toby. I would also like to thank Rahim Shayegan and Yona Sabar, my dissertation committee members, for their encouragement and support. I utterly absolve these generous mentors of responsibility for any errors in this study.

I would like to devote a special paragraph of thanks to the School of Abbasid Studies, whose open, encouraging, and invigorating atmosphere makes me feel sure that I am in the right profession. I thank all of its members, including Devin Stewart (for guidance on questions of Shīʿism), Paul Heck (for an e-mail on theo-humanism), John Turner (for a life-changing correspondence that is currently being converted into an article), and James Montgomery (for the 'cookbook'). I would also like to thank everyone who engaged me in conversations on the subject of the *Ḥikāya* at the Istanbul meeting in August 2014. I would especially like to thank Julia Bray and Wen-Chin Ouyang for their eloquent and insightful comments on my paper, and for their support and careful attention as editors of the Classical Arabic Literature series for Edinburgh University Press.

I would like to thank my colleagues and friends in Manchester, including Peter Pormann (for teaching me to fear a lack of philological rigour), and Taro Mimura (for sharing an office with me whilst radiating calmness and goodwill). I thank Stevie Spiegl, to whom I literally owe my life, so often has he fed me dinner. I thank Jak Kenton-Spraggan for teaching me the punk rock ethos, and thanks to whom I am now myself a punk rocker. I thank Gwen Gawthrop for showing me that when you think you cannot run any farther, you are just getting started, and I thank Joe Walker, for talking to a stranger in the park. I also thank Micheal Kingsley of Teignmouth for helping me think of a synonym for 'epically-described'. I would like to dedicate this volume to the city of Manchester, for making a wanderer feel at home.

Finally, I would like to thank my beloved family, Deborah and Peter Selove (my parents), and Matthew and Benjamin Selove (my brothers), whom I have missed very much since I moved abroad. Special thanks to my mother for providing me with publication information for a volume of *Winnie-the-Pooh,* and for reading to me from this same volume when I was little.

The Ph.D. dissertation on which this book was based was made possible with funding from the University of California Institute for Humanities Research, the UCLA Center for Medieval and Renaissance Studies, the UCLA Graduate Division, the Mellon pre-dissertation fellowship, Harry and Yvonne Lenart fellowship, and the UCLA Dean's fellowship. It was revised into a book while I was a research associate at the University of Manchester.

Cover Illustration Acknowledgements and Explanation

The painting on the cover of this book was based on a photograph of many of my dear friends in Manchester. I am grateful to them for dressing up in bathrobes and false beards for me. Although this painting is based loosely on the style of illustrations of al-Ḥarīrī's *Maqāmāt* (a work roughly contemporary with the *Ḥikāya*), you will notice several anachronistic elements: especially a modern-looking beer bottle, spectacles, and fingerless leather gloves. By including these features I meant to suggest something about our inability to date this text with any certainty, a synecdoche for the general uncertainty surrounding a work that nevertheless continues to amuse and engage us probably a thousand years after its composition. Each person represents a character at the party that Abū al-Qāsim crashes, listed here starting at the left side of the painting. Matthew Swarbrick represents 'a person who likes to joke and banter' (*insān yamzaḥ wa-yataṭāyab*).[1] Gwen Gawthrop (seated to his right) is 'a silent person at the party' (*sākit fī al-majlis*).[2] The two of them together also represent 'two friends' (*ṣādiqān*).[3] Seated to her right is Gabor Gergely, representing a man with a large turban.[4] He was originally meant to be a 'very important person' (*insān khaṭīr*)[5] until I realised that this referred to the same character represented by Luca Larpi, the writer or secretary (*al-kātib*),[6] seated to his right. Stevie Spiegl, standing in a pink bathrobe, represents his servant, described only as 'the standing black man' (*al-aswad al-qā'im*).[7] To his right stands Aaron Marshall, the host's butler (*wakīl ṣāḥib al-dār*).[8] Seated in the middle of the painting, in all his glory, is Jak Kenton-Spraggan, the host of the party. The two musicians in the front are Hannah Cawthorne (on the tanbur), and Peter Maskell (on the lute).[9] Behind them sits Cat Naylor, the 'beardless youth' (*amrad*).[10] Seated from left to right behind her are Joe Silver, the 'tough guy' (*jald*),[11] Taro Mimura, the 'very learned man' (*rajul fāḍil adīb*),[12] and Will Ward, the 'frequent visitor of important people' (*insān yudākhil kibār*).[13] Standing behind them are Helen Spencer, the 'singing girl' (*al-mughanniyya*),[14] and Mark Snowden, her guard (*raqīb*).[15] Abū al-Qāsim's body is based on an illustration of al-Ḥarīrī's *Maqāmāt,* but the face is a poor copy of Norman Funicello's (1942–2007). For some reason it came out looking more like Stephen Cummins', Norman's (and my) dear friend, and owner of Indian Creek Farm in Ithaca, New York. I am in the front of the painting, representing the Daylamī slave boy, serving my friends and my dear readers as humbly as I may. The table is based loosely on medieval Arabic maps of the world, and represents the microcosm.

Notes

1. *Ḥ*, p. 70.
2 *Ḥ*, p. 88.
3 *Ḥ*, p. 81.
4 *Ḥ*, p. 62.
5 *Ḥ*, p. 59.
6 *Ḥ*, p. 58.
7 *Ḥ*, p. 61.
8 *Ḥ*, p. 76.
9 *Ḥ*, pp. 67, 366.
10 *Ḥ*, p. 78.
11 *Ḥ*, p. 55.
12 *Ḥ*, p. 56.
13 *Ḥ*, p. 63.
14 *Ḥ*, p. 336.
15 *Ḥ*, p. 337.

Introduction

> This Right Whale I take to have been a Stoic; the Sperm Whale, a Platonian, who might have taken up Spinoza in his latter years.
>
> Melville, *Moby-Dick*, 'The Right Whale's Head – Contrasted View'

Ḥikāyat Abī al-Qāsim, written by the otherwise unknown al-Azdī, probably in the eleventh century AD, describes a party in Isfahan that begins in the morning and lasts well into the night. Nothing impossible happens at this party: a group of fairly important, decent people have gathered together, food is served which is good but not extraordinary, and capable (though not legendary) entertainers and servants cater to the guests. The ordinariness of the event, however, is overwhelmed by the presence of a remarkable and uninvited guest, Abū al-Qāsim Aḥmad ibn ʿAlī al-Tamīmī al-Baghdādī, who dominates the conversation with his wide-ranging, prolix discourse, which he spews in a quantity just barely possible for an actually human old man who had long earned his bread by entertaining with his conversation, which is who Abū al-Qāsim seems to be.

In his introduction to the *Ḥikāya*, al-Azdī explains that the events he represents occupy a day and a night, and can be read in that same amount of time. He also tells us that the Baghdadi guest, who dominates the conversation with his alternately obscene and elegant tirades, represents the entirety of Baghdad, and indeed a kind of microcosm of creation. The story itself is narrated in a past-continuous, iterative tense, as in 'Abū al-Qāsim would walk into a party', or 'Abū al-Qāsim would say', for example. Thus his act of party-crashing is described only as an example of the kind of thing that Abū al-Qāsim might typically do.[1]

Tempted by al-Azdī's promise to provide a microcosm of his

contemporary Baghdad, readers may look to the *Ḥikāya* for an example of realism, hoping to find within its pages a true-to-life microcosm of the eleventh-century city. Although the *Ḥikāya* brims with words for household furniture and food, we may be disappointed in our quest for a realistic depiction of the everyday, which is seemingly marred by the overabundance and obscenity of Abū al-Qāsim's speech. Abū al-Qāsim tells us things about Baghdad we may find nowhere else in literature, describing the sounds of the water-wheels in the river, a story about excrement in the streets, and Baghdadi swimming strokes with names like 'the scorpion' and 'the peacock', named but not defined.[2] For just at the moment it seems that Abū al-Qāsim's discourse brings the physical presence of the city closer than ever before, at the next moment his speech seems, like Baghdad itself, a mere literary figment, elaborately imagined.

The title alone, difficult to translate, seems to hover between the promise of a realistic portrait and a grotesque parody. The word *ḥikāya*, as in the title *Ḥikāyat Abī al-Qāsim* could mean an imitation or mimicry, or it could mean a story.[3] However, even if al-Azdī is providing, as he seems to promise, an imitation or mimicry of contemporary Baghdad or Baghdadi speech, mimicry itself can imply not necessarily a faithful reproduction of the thing imitated, but a playful exaggeration.

A mixture of the manneristic and the realistic, the story unfolds in a fashion seemingly unique in the history of classical Arabic literature, comprising a continuous narrative in a single setting. Only by the names and places mentioned in the text itself were Adam Mez and others able to deduce that the *Ḥikāya* was probably composed in the early-eleventh century, since which it disappeared without a trace in the documentation, surviving only in a single manuscript now held by the British Library in London. Mez produced the first edition of this manuscript with an introduction in German in 1902, exciting the opprobrium of his fellow scholars, one of whom declared that it was 'so disgusting a text to be unworthy of a serious scholar's attention'.[4] Despite its often aggressive obscenity and blasphemy, the text nevertheless has been hailed by all, even its detractors, as a unique innovation in Arabic literature.[5]

What is the nature of this text, at once a collection of quotations from other works, and a highly unusual innovation? As the author tells us in his introduction, it is a microcosm of the city of Baghdad, of literature, or of humankind. The text seems to strive to contain everything and its opposite in one day's worth of conversation, inviting close readings and philological investigations with its outlandish, puzzle-like language, as well as offering doorways to the broader world of satire and banquet literature as it adopts and explores tropes of food and wine-consumption, party-crashing, and invective that echo throughout ancient and medieval literary traditions.

Manuscript

The unique codex manuscript, held in the British Library, contains 132 folios with about fifteen lines per side, including a title page with the full title, *Ḥikāyat Abī al-Qāsim al-Baghdādī al-Tamīmī wa-l-ʿajāyib wa-l-gharāyib ʿalā mā jamaʿat min al-ḥikāyāt* (The Imitation of Abū al-Qāsim al-Tamīmī the Baghdadi, and the strange and wondrous things according to the imitations therewith collected). It is written on thick, high-quality paper with margins of about ½ inch, evidently trimmed at some point. The spine of the British Library binding reads 'Abu 'L-Mutahhar Muhammad al-Ardi [sic], HIKAYAT ABI L-KASIM AL-BAGHDADI'.

The history of this undated manuscript is obscure, but the title page informs us that it was part of the library of Ṣāliḥ ibn Muḥammad ʿAbd al-Laṭīf (whose identity as yet remains a mystery). A piece of white tape on the first folio covers the name of a former owner, which can be read by shining a light through the back of the paper. Underneath the tape, we find 'Ex Libris Theodor Pres. S.S. Trin'. This appears to refer to Theodore Preston, Lord Almoner's Professor of Arabic during 1854–71 in Trinity College, Cambridge, who studied and translated the *Maqāmāt* of al-Ḥarīrī, and likely was interested in the *Ḥikāya* because of its resemblance to *maqāmāt*-style stories of eloquent tricksters. The final folio of the manuscript is marked with 'purchased at Sotheby's, 3 August 1854, lot 708', and indeed the Sotheby's auction record for that year lists 'The Comic Tales and Anecdotes of *Abou'l*, Kasem of Bagdad [sic]', part of the collection of John de Whelpdale, Esq., of Armathwaite, d. 1844, whose manuscript and engraving collection was auctioned off ten years after his death. Since practically nothing is known about the transmission of this text, further research into Theodore Preston's and John de Whelpdale's manuscript collections may prove fruitful in reconstructing the history of the *Ḥikāya*.

The manuscript is written in a tidy and well-vowelled *naskh* script, with Coptic lettering suggesting that it originated in Egypt.[6] A marginal note on folio 83a tells us that the manuscript was read in Shawwāl 727 AH (August 1347), so it must have originated before that date. Despite the neat and well-vowelled script, it contains numerous inscrutable words and phrases: a few may be scribal errors, but most are probably colloquialisms or specialised terms whose meaning can only be guessed at.[7] The scribe seemed particularly uncomfortable with the few Persian colloquialisms, which he frequently leaves undotted, and which are especially difficult to decipher.

Authorship and Dating

In the opening lines of the manuscript, the author is identified as the otherwise unknown Abū al-Muṭahhar Muḥammad ibn Aḥmad al-Azdī. Because no mention of this author or his undated text can be definitively located in any contemporary biographical dictionaries, virtually every modern scholar of the *Ḥikāya* has addressed the question of its dating and authorship, a question debated even in the opinion section of *al-Ḥayāt* newspaper.[8] Based on the names and places mentioned in the text, all agree it was probably composed in the early-eleventh century. Mez, the first to make these calculations, notes that the narrative includes such figures as Abū Naṣr ʿAbd al-ʿAzīz ibn Nubāta (d. 1014) and Ibn Ghaylān al-Bazzāz (d. 1048), but does not mention any changes to the city of Baghdad associated with the Seljuq empire's influence on that city, the beginning of which (c. 1055), he provides as the *terminus ad quem.*[9]

Most of the debates concerning its authorship centre on ʿAbbūd al-Shāljī's decision to edit the text as a lost work of Abū Ḥayyān al-Tawḥīdī, *al-Risālat al-Baghdādiyya* (and indeed the author, whoever he may be, refers to his work as a *risāla*, or epistle, at the end of his narration).[10] ʿAbd al-Laṭīf al-Rāwī, in his article 'A-hiya *al-Risālat al-Baghdādiyya* am *Ḥikāyat Abī al-Qāsim al-Baghdādī*?', argues that the *Ḥikāya* was instead written by the poet Ibn al-Ḥajjāj, who was born in Baghdad c. 941, and was well-known for his obscene and scatological verses, often adopting the language of the streets.[11] Like al-Tawḥīdī, he is quoted extensively in the text. Nevertheless, these extensive quotations are practically the only basis for either argument, and given that the *Ḥikāya* contains hundreds of quotations from as many sources, and given, furthermore, that we have only one manuscript from which to derive information, any argument about the *Ḥikāya*'s authorship can only be speculative. Since this problem is truly unsolvable with the information currently available, I will refer to the author of the text as al-Azdī, as he is named on the manuscript.[12]

Editions

Adam Mez was the first to edit the manuscript. In 1902, when few scholars deemed obscene material worthy of their study, he was well ahead of his time. Acknowledged as one of the first scholars to separate the study of Semitic languages from the study of theology,[13] Mez was interested in the text as a source of cultural history, and in its protagonist, Abū al-Qāsim, as a *Sittenbild*, or depiction of the manners and customs of a typical Baghdadi (his edition is titled *Abulḳāsim, ein bagdâder Sittenbild*). In his introduction, he situates the text's emergence within the history of Arabic literature, particularly the development of urban entertainment literature, and its 'discovery of empirical man'.[14] He devotes special attention to the literary innovators Abū ʿUthmān ʿAmr ibn Baḥr al-Jāḥiẓ (d. 868), and Ibn al-Ḥajjāj (d. 1001), both of whom al-Azdī cites copiously in his introduction. Al-Jāḥiẓ, born in Basra, is often called 'the father of Arabic prose', and is considered the first to have written about certain 'lowly' everyday topics not previously typical to literature, such as misers and party-crashers. Ibn al-Ḥajjāj is mentioned above in the Authorship and Dating section, as his obscene style is so close to that of the *Ḥikāya* that he was proposed as its possible author.[15]

Mez's contemporaries mostly scorned his efforts as misdirected towards obscene filth, and indeed his harshest reviewer, M. J. de Goeje, confessed himself to have been unable to read the entire story, which he found disgusting and low: '*Viel Schmutz, wenig Geist*'.[16] His review nevertheless provides a very lengthy and useful list of emendations to the problematic edition. Carl Brockelmann also provides a small list of emendations, and also expresses some disgust. He praises the innovative style of Mez's unique subject of study, but blames the *Ḥikāya*'s obscene and difficult vocabulary for the lack of imitators of its newly invented *Sittenbild* form.[17]

In 1980 ʿAbbūd al-Shāljī produced a far more readable edition of this difficult text, publishing the *Ḥikāya* under the title of a lost work of Abū

Ḥayyān al-Tawḥīdī (d. 1023), *al-Risālat al-Baghdādiyya*. He dedicates most of his introduction to a biography of this well-known scholar and writer, who has in common with the *Ḥikāya*'s protagonist a disagreeable personality and a hurtfully sharp tongue, as well as a propensity for wandering and an affinity for mystical religious practices.[18] Though al-Tawḥīdī is quoted extensively in the *Ḥikāya* and shares certain similarities with its protagonist, he cannot be regarded as its author with any certainty, and thus al-Shāljī's lengthy biography and critique of his writings can only be considered tangentially related to the text itself.

Himself a Baghdadi, a cultural historian, and an editor of other similar texts, al-Shāljī is able to recognise many (but by no means all) of the obscure idioms and names of foods and goods found in the manuscript. Al-Shāljī's edition shows considerable improvement over Mez's, whom he praises for his efforts in editing such a difficult manuscript. He is himself forced to leave countless words and idioms largely unexplained, or only tentatively interpreted. Though he often notes his amendments to the manuscript, on rare occasion he makes substantial changes to a word or phrase without any note or justification.[19] His edition also provides an index of foods and goods that are mentioned in the story.[20] Such an index would prove useful to scholars seeking to use this work as an encyclopedia of material culture, which some have done (as shown below).[21]

Translations

There have been two translations of the manuscript. Mary St. Germain translates it into English in her unpublished dissertation, Placing an Anomalous Text within the Literary Developments of its Time (2006), in which the story is rendered in literal, non-literary, academic language. There is also a loose, popularising French translation by René Khawam, *24 heures de la vie d'une canaille* (1998). St. Germain, in her introduction, provides a detailed review of the debate surrounding the authorship and dating of the text, and describes the passages quoted and features shared with other contemporary Arabic works, such as the *Maqāmāt* of al-Hamadhānī and various works of al-Tawḥīdī. She then argues that the *Ḥikāya* was a subversive prose reworking of the pre-Islamic *qaṣīda*, a typically long poem that begins with a lament at the abandoned campsight of a beloved, followed by a description of the poet's subsequent travels through the desert, and ends with a section of praise or satire delivered at the end of these travels. St. Germain reads the *Ḥikāya* as deliberately subverting each of these sections throughout the course of the day it depicts.

Addressing a much more general audience, Khawam introduces the *Ḥikāya* in a classically orientalist style, claiming it as proof that medieval Islam was *le lieu géométrique de toutes les libertés* (the locus of all liberties) and marketing his translation as sensationally obscene. He divides the text into chapters based on the hours of the day, a creative innovation based, perhaps, on al-Azdī's introduction, in which he states that the events described occupy a single day and night. Thus, with St. Germain's attention to the *qaṣīda*-like structure of the narrative's movement, and Khawam's imaginative chapter divisions, both translators attempt to account for the unusual structure of the text, and both representations will find a place in my own discussion of the question of narrative structure in the *Ḥikāya*. Since this structure is seemingly unique in Arabic literature, this question unavoidably presents itself to any thorough reader of the *Ḥikāya*, and no one is forced to read more carefully than a translator. I am also co-producing an edition and translation of the manuscript (on which see more below in Chapter 1).[22] This translation will attempt a compromise between St. Germain's literal approach and Khawam's popularising style. However, given the nature of the manuscript, which, though neatly written, brims with hapax legomena and errors of a confused scribe, any translation, or indeed edition at this stage, must necessarily be considered provisional.

Scholarship

Modern scholarship on the *Ḥikāya* began with Mez's edition of the text and responses to it. Mez's and these other early treatments of the work often compared it to classical texts, and, focusing on the word *ḥikāya* as an Arabic translation of mimesis, described it as a (failed) work of realism. In seeking a materiality outside the text, some subsequent scholars also used it as an encyclopaedia of material culture. More recent treatments of the *Ḥikāya*, beginning with Abdelfattah Kilito's in *Les Séances* (1983), have taken a more deconstructive approach, focusing on the deceptive and disorienting qualities of its language. They have largely ceased comparing the work to its classical equivalents, as indeed such comparisons, once conducted in a fashion that now seems outdated, held up the classical texts as the standard which the *Ḥikāya* ultimately failed to meet.

Mez, first editor of *Ḥikāyat Abī al-Qāsim,* was interested in obscene literature not only as a development of classical genres but also as a possible source of historical information. For example, he based his cultural encyclopaedia of Buyid Iraq, *The Renaissance of Islam*, partly on such works of Arabic literature.[23] The Buyid dynasty, which lasted from the mid-tenth century until 1055, was Persian and Shīʿī, and although the Buyids made a show of recognising the authority of the figurehead Sunni Abbasid caliphs, and although they were active patrons of Arabic writers and poets, their reign fostered tension between Persian and Arabic-speakers, and between Sunnis and Shīʿīs. It could be argued that this tension can be felt in the *Ḥikāya*, and thus it does reflect its physical environment.

In Mez's introduction to his edition of the *Ḥikāya*, he writes that its author concerned himself not with 'the concept' (*Begriff*) but 'the individual thing' (*einzelne Ding*),[24] and was inspired not by rhetoric but by observation, taking pleasure in the real (*Freude an den Realien*). European scholars and writers, especially from the 1830s onward, displayed particular interest in the 'realism' of literature. This interest was part of the modern reception of ancient notions of mimesis.[25] Mez's introduction to the *Ḥikāya* is clearly influenced by these scholarly projects and by the interests of his time.

Mez was a pioneer in the field for bringing Arabic literature into the conversation about the reception of classical styles and methods, for scholars had previously studied the Arabic language in conjunction with Biblical Hebrew, and, in Germany, often in the faculties of theology.[26] Mez, however, approached the *Ḥikāya* and other works of Arabic literature in the tradition of scholarship on works of ancient Greek and Latin, referring in his introduction to parts of the narrative as *das Symposion* and *das Satyrspiel.* Early scholarship on the *Ḥikāya*, all of which is dependent on Mez's innovative work, returns again and again to questions of its relationship to works of classical literature, and to its depiction of 'the real'.

Part of this discussion concerns translations of the word *ḥikāya* as 'mimesis', a question that also relates to the work's potential realism. Duncan B. Macdonald notes that Aristotle's 'mimesis' was sometimes translated into

Arabic as *ḥikāya*, and that 'the conception of literary art as an "imitation" of life may thus, when translated into Arabic forms, easily have resulted in Abū 'l-Muṭahhar's new literary type.'[27] In recounting the history of literary developments preceding the composition of the *Ḥikāya*, Mez emphasises the role of 'imitation with comic exaggeration',[28] as well as of 'Mimesis', which could have a comic effect, but which could also help ensure reliable transmissions of the sayings and deeds of the prophet Muḥammad (hadiths), which had legal significance. Soon after Mez's edition of the *Ḥikāya* appeared, Josef Horovitz published his *Spuren griechischer Mimen im Orient*, which traces elements of Greek mime in early Arabic popular culture and literature (1905).[29] Horovitz, who employs 'mime' as in 'farce' interchangeably with 'mime' as in 'imitation' when discussing the *Ḥikāya*, uses this work and Mez's introduction as a source for describing the role of mimicry in Arabic literature and culture, both concerned with word-for-word transmission of hadiths and *akhbār* (anecdotes or reports), and with the dialectical peculiarities that distinguished tribes and ethnic groups.[30]

Such comparisons between Arabic and the Classics inevitably implied a hierarchy of value, and Greek and Roman literature were the standard against which Arabic literature was judged. For example, Gustave von Grunebaum (who also describes the *maqāmāt*, a type of literature intimately related to the *Ḥikāya*, as 'the attenuated offspring of the classical *mimoi*'),[31] writes in his account of Greek influences on the *1001 Nights* (1946):

> The Arab was not accustomed to that historical narrative in which some of the romances excel, and he had, on the whole, no experience in inventing and carrying through a complicated action, with many secondary actions to boot, stretching over hundreds of pages. These differences in literary tradition make for a loss of refinement, greater simplicity, or, perhaps, obviousness of the Arabic tales …

Because of these shortcomings of the Arabic borrowings, which he ascribes largely to their transmission by 'professional story-tellers' as opposed to 'professional writer-rhetoricians … The artistic level is bound to drop'.[32] Here von Grunebaum counts not only the Arabic literary tradition as inferior to the Greek, but also oral popular traditions of the 'professional story-teller' as inferior to elite written narratives.

The author of the *Ḥikāya* almost invites such criticisms with his introduction to his work, in which he apologises for his low or faulty language (*laḥn*) and expresses concern that he will be perceived as having a lack of knowledge (*quṣūr maʿrifa*).[33] Without acknowledging this apology, early scholarly criticisms of the *Ḥikāya* do tend to find fault with its language (described as too obscene or colloquial), and accuse it of lacking balance and unity as compared to 'high' literature. Even Mez criticises the *Ḥikāya* as a 'bungling first attempt' at a style new to Arabic literature, and dependent on Greek *synkrisis* (literary comparisons).[34] Horovitz, in his *Encyclopaedia of Islam* entry on al-Azdī (1913) expresses disappointment in the text, claiming that

> the author, led on by his philological inclinations, has interwoven so much of his extensive knowledge of the *adab* literature and terminology of the different trades and also of pornographic poetry ... that the realism of the description as well as the unity of the tale suffer considerably.[35]

By 'realism' Horovitz here appears to mean 'plausibility', implying that the conversation portrayed in the *Ḥikāya* is too overfull and abundant to be humanly possible. However it is an overfullness of another type of 'realism', language based on the jargon of traders, or language earthy in its subject matter, that is charged as being responsible for its very lack of realism.[36]

A half-century afterwards, the *Ḥikāya*'s disorienting language continued to earn it both criticism and unfavourable comparisons to the classics. One of the earliest articles written on the *Ḥikāya*, Francesco Gabrieli's 'Sulla *Ḥikāyat Abī al-Qāsim* di Abū l-Muṭahhar al-Azdī' (1942), describes the 'curious paradox' that al-Azdī's 'daring attempt at realism' was seemingly derived from an Aristotelian concept of *universal* mimesis,[37] alluding to the tension between peculiarity and generality presented by the text. He goes on to paint the *Ḥikāya* as a 'bud of the rare plant of Greek mime' (*un germoglio della rara pianta del mimo greco*), which, however, failed to bloom due to the overly luxurious literary tastes of its day.[38] In Gabrieli's case, such criticisms were probably founded on his adherence to the 'idealistic aesthetics' of Benedetto Croce, which sometimes led him 'polemically' to champion 'the right of Western scholars to examine and judge Arabic literature according to universal aesthetic criteria'.[39]

Gabrieli further criticises the *Ḥikāya*, characterising it as al-Azdī's failure to 'remain Petronius so as not to become Athenaeus'.[40] By this he implies that instead of providing a humorous, readable narrative like the *Satyrica* of Petronius,[41] the *Ḥikāya* instead provides a tedious overabundance of literary citations and strange words. And in that regard, he suggests, it resembles Athenaues's *Deipnosophistae* (third century AD), a lengthy Greek work about a dinner conversation, encyclopedic in scope, and describable as dull. James Davidson, however, writes of the dullness of the *Deipnosophistae*:

> This relationship between a titillating subject-matter and what may be considered to be a dull, pedantic text which zooms in constantly on mere words is not . . . explicable in terms of accident or of a failure on the part of the author. It is interesting in its own right and needs to be understood as absolutely central to the whole work.[42]

I discuss the function of boredom in the *Ḥikāya* in 'Those Camels have Passed' in Chapter 2,[43] and, like Davidson, I see it not as a failure of the author, but as a part of this microcosmic work that deserves its own analysis.

Though this study is dependent on the insightful work of its early scholars, any value judgments that they made in describing the *Ḥikāya* naturally strike us as misplaced from our post-modern perspective, which instead values the text as a world in itself. Thus, some of the faults that Mez finds with the narrative, for example, that the section on horses is 'too long',[44] and that the supposedly Baghdadi Abū al-Qāsim at one point (as if accidentally) claims to be from Isfahan, will form part of my analysis of this text, which takes them not as mistakes as Mez does, but rather as part of the ecosystem of this microcosmic work. Similarly, while Horovitz criticises the overabundance of the protagonist's speech as detracting from the text's realism, this same overabundant speech is central to my analysis of the *Ḥikāya*'s disorienting use of language. As Stephen Halliwell writes,

> ancient mimeticism moved around, and was partly energised by the tension between, two major poles of thought, one a sense of mimesis as a reflection of and engagement with 'external' reality, the other an inclination to think of mimesis as the creation or invention of self-contained, fictional worlds.[45]

I read the *Ḥikāya* (which I would translate as the *Imitation*) as this second form of mimesis, though I read it as having a manneristic twist, in that it seems to question our daily perception of reality with its language-based creation of a second world. Founded on an ancient reading of mimesis, my analysis is nevertheless more in tune with the modern deconstructive reading of Abdelfattah Kilito (1983), and thus returns to comparing it with classical texts, but in a new interpretive light. Kilito's work is described below,[46] following an overview of other earlier works of scholarship which are here presented chronologically.

After Mez and the contemporary European scholars who responded to his edition, Zakī Mubārak was the first scholar to address the *Ḥikāya*, devoting a chapter to the work in his anthology of eleventh-century Arabic prose, *al-Nathr al-fannī fī al-qarn al-rābiʿ*, his Ph.D. thesis at the Sorbonne (1931).[47] Mubārak, in reference to the author's introductory apology for his work, writes that ungrammatical Arabic can prove funnier than high literary eloquence (*al-laḥn qad yakūn amzaḥ min al-faṣāḥa*).[48] According to Mubārak, however, because of this language, the author cannot keep his promise to provide a portrait of all Baghdadis, high and low, but actually provides a portrait of only one side of the city, the silly side (*waṣf jānib khāṣṣ huwa jānib al-ʿabath wa-l-mujūn*).[49] Thus, like Horovitz, he contends that this narrative's low language actually detracts from the 'realism' of its portrayal of the city of Baghdad. The preponderance of filthy language in this work, however, Mubārak reads as a reflection of the filthiness of an urban environment, with its odours of the fish markets and the corpses of animals.[50] Thus, he sees the *Ḥikāya* as a function of the urban environment from which it appears to have originated,[51] but disproportionally influenced by certain 'low' aspects of this environment.

In his article, 'Aspects of Arabic Urban Literature' (1955),[52] von Grunebaum also sees the *Ḥikāya* as a product of its contemporary urban environment. In making this claim, he compares al-Azdī to Petronius:

> [Each] holds up to his contemporaries a realistic caricature of the ways of the town and its polite society, with their absurdities and their elegance, their beauty and obscenity and, above all, of that incessant effort to make the repulsive aspects of our existence less hurtful to human pride and to stylise our lives into compatibility with that heightened concept of human dignity whose actualisation Islam imposed on urban society as its foremost task.[53]

Here von Grunebaum characterises the *Ḥikāya* as a 'realistic caricature' (possibly an oxymoron). He is struggling, like his predecessors, to describe what kind of portrait of Baghdad this work presents. It is realistic, he suggests, in that it challenges constructions of identity that people use to hide their own absurdity from themselves. By likening al-Azdī to Petronius in this regard, he refers, perhaps, to the latter's parodying of epic literature (a subject of discussion in the Conclusion of this present study). Von Grunebaum's earlier assertion in his article that these texts constitute a 'discovery of empirical man'[54] echoes Mez's discussion of the *Entdeckung des empirischen Menschen*[55] in the development of Arabic prose genres that preceded the composition of the *Ḥikāya*. Although the import of this phrase as used by Mez and by von Grunebaum differs somewhat, both are interested in the representation of the particular as opposed to the general, or even to an ideal form.

Despite the various complexities and difficulties found in discussions of the *Ḥikāya*'s realism, some sources use it as an encyclopedia of material goods available in eleventh-century Baghdad. ʿAbd al-Wāḥid Dhū al-Nūn Ṭāhā in his article 'Mujtamaʿ Baghdād min khilāl *Ḥikāyat Abī al-Qāsim al-Baghdādī*' ('Baghdadi Society through the Lens of *Ḥikāyat Abī al-Qāsim al-Baghdādī*', 1974) recognises the *Ḥikāya* as a work of *mujūn* (licentious absurdity),[56] and says it is too ugly to genuinely portray the wide range of classes and social levels promised by the author in his introduction. He supports this assertion with a quotation from Zakī Mubārak, described above.[57] Nevertheless, Ṭāhā goes on to use the *Ḥikāya* as the encyclopedia of Baghdadi foods and material goods that it sometimes appears to be.

Likewise, in his *Social Life under the Abbasids* (1979), Muhammad Ahsan briefly introduces the *Ḥikāya* as 'a satirical but realistic picture of life and manners in Baghdad'[58] (not unlike von Grunebaum's contradictory description of the text as a 'realistic caricature'). Ahsan then draws from the *Ḥikāya* and other literary texts in a largely uncritical fashion to provide information about the housing, clothing, and foods of Abbasid times. For example, he uses the *Ḥikāya* as his sole source for his claim that 'In the houses of the common and less cultured people, the guests were taken [to wash their hands after dinner] to a drain (*bālūʿa*) flowing at the end of the courtyard of the house.'[59] The source for this claim is a viciously satirical passage, which, because of its tone, surely should not be considered a reliable source of information about medieval hand-washing practices:

> A bumpkin comes out, a middle-aged yokel from Sawād[60] with the physique of a camel and a thick grey beard, poor and threadbare, with sticks of firewood in his hand. He gives them out as toothpicks, then herds everyone into the courtyard to wash their hands in the drain, which breaks, by God, the noses with the smell of all the garbage collected in it. May God burn this species of manhood![61]

Similarly, Shelomo Dov Goitein in his *Mediterranean Society*, writes,

> I have always been puzzled by the prominence of cushions in the outfits of a bride. They were, of course, an intrinsic part of the seating and sleeping facilities, but their full function was revealed to me when I came upon a passage in Abu 'l-Qasim's *Ḥikāya*. He writes that a house full of cushions looks like a ground covered with flowers. One type of cushion in the Geniza is indeed called 'garden pillow' . . .[62]

However, the list from which Goitein draws his information about pillows is actually a list of things Abū al-Qāsim does not see in Isfahan, and in which he is apparently imagining the most luxurious array of pillows possible in order to make the Isfahanis feel inadequate:

> I do not see your houses spread with carpets of al-Maghrib, Anatolian mats, rugs of Andalusia, or Cordoba, or of the regions of Armenia. Or Byzantine hair mats, sitting mats of Tustar, leather rugs, Maghribi gold-embroidered leather mats, pillows of Dabīqī gold, Cyprus sitting throws, Sūsanjirdī tapestry-coloured cloth, iridescent carpets, or floor cushions that make the room seem like a field strewn with blossoms.[63]

As is shown throughout the present study, it is difficult to trust the *Ḥikāya* wholeheartedly as a source of such material information, especially when it is provided in lists of things that the trickster speaker does not see.

However a sub-genre in medieval Arabic literature that focuses on beggars or other marginal figures does often style itself as a kind of realism. For example, al-Jāḥiẓ's introduction to his *Kitāb al-bukhalāʾ* (*Book of Misers*) compares this work to another on the tricks of thieves (not extant), in being not only entertaining but genuinely useful in understanding and avoiding the undesirable habits of these types of people.[64] This is more or less in tune with readings of such literature as portrayals of the filth (literal and metaphorical) of an urban environment. In *The Mediaeval Islamic Underworld* (1976), Clifford Edmund Bosworth attempts to situate medieval Arabic literary portrayals of low-life groups and their secret jargon in their historical and cultural contexts. He focuses on Abū Dulaf's tenth-century *Qaṣīda sāsāniyya*, a lengthy poem complete with commentary by the author and filled with jargon of the *Bānū Sāsān*, wandering rogues who employed trickery to earn their bread. 'The interest in low life and all its manifestations', he writes, 'so characteristic of the 3rd/9th and 4th/10th centuries, seems to stem from the progress of urbanisation in the Islamic lands, with the concomitant economic prosperity making people more aware of the gaps between social classes.'[65] Giving a brief description of the contents of the *Ḥikāya*, whose protagonist in many ways resembles the wandering writer Abū Dulaf himself, Bosworth characterises the text as 'the culmination of nearly two centuries' evolution of this urban poetic development'.[66]

The author of the *Ḥikāya* does offer his work as a sample of the speech of the Baghdadis, and one may plausibly ascribe the obscurity of its language at least in part to its use of a particular urban jargon. Furthermore, the long lists of foods and trade goods may indeed prove obscure to the modern reader due to their highly local and time-restricted specificity. Nevertheless, readings of the *Ḥikāya* as a work of realism depicting eleventh-century Baghdad must be tempered with an awareness of its use of language in a way that sometimes seems to call reality itself into doubt. Kilito's *Les Séances*,[67] an abstract, evocatively fanciful, and astute exploration of the *maqāma* genre, with which the *Ḥikāya* is often compared, evinces keen awareness of the deceitful qualities of language in this text. In his section devoted to the *Ḥikāya*, Kilito explores the idea of opposition embodied in the person of Abū al-Qāsim, whom he compares to a *ḍidd*, a word that means both a thing and its opposite. With this comparison, Kilito not only characterises Abū al-Qāsim as inextricably aligned with the vagaries of language, but also avoids the temptation to identify Abū al-Qāsim's roguish persona as his true personality, and his pious persona as a mere disguise.[68]

Kilito also emphasises the idea of cyclically reoccurring time in the

Ḥikāya, an idea fruitfully elaborated in a later article by Philip Kennedy, 'The Maqāmāt as a Nexus of Interests'.[69] This article situates the *Ḥikāya* with respect to the *maqāmāt* genre, arguing that a *ḥikāya* is distinct from a *maqāma* in its mimetic, imitative aim.[70] Like Kilito, he focuses on the *Maqāmāt* of al-Hamadhānī (d. 1008), which tell a series of stories in which the narrator, ʿĪsā ibn Hishām, repeatedly fails to recognise the protean trickster Abū al-Fatḥ al-Iskandarī as he delivers speeches, preaches, and begs under many disguises. Kennedy opens his study with a definition of the *maqāma* itself (important to any analysis of the *Ḥikāya*) as 'a picaresque anecdote written in rhyming prose (*saj*) which crystallized into a genre in the late [...] 10th century AD', in which, '*adab* literature (educative collections of anecdotes, usually involving historically identifiable personages) is [...] fashioned into a self-consiously rhetorical form in which features of language and style can seem to outweigh the narrative content.'[71] As in the *Maqāmāt*, he writes, we can discern in the *Ḥikāya* 'the conscious development of the idea of fiction which causes anxiety in *adab*, and which *adab* seeks to avoid'.[72] Kennedy vividly describes the *Ḥikāya* itself as a kind of 'descent into madness', emphasising Abū al-Qāsim's 'altered state'.[73] He also raises the question of the reader's experience of its sometimes tedious portrayal of the passage of time.[74] These features of the *Ḥikāya* highlighted by Kennedy and Kilito are the subject of extensive exploration in this present study, and especially in Chapter 4,[75] which deals with the ambiguity of its language, and in 'Those Camels have Passed' in Chapter 2,[76] which describes its representation of time passing.

Shmuel Moreh's controversial *Live Theatre and Dramatic Literature in the Medieval Arabic World* (1992) contains an original attempt to wrestle with the *Ḥikāya*'s unusual portrayal of the passage of time.[77] Moreh sees the *Ḥikāya* as a script for a play, whose over abundant language is meant only as material inspiring to the improvisational actor. As his critics have already pointed out, his theory that the *Ḥikāya* was the script for a live, improvised performance is probably indefensible. This theory was based in part on his reading of the text's iterative-style narrative as an invitation to improvise, as all the actions described are presented only as hypothetical examples of actions that might have been performed. Everett Rowson, however, suggests in his review of Moreh's work that this style of narrative serves, in fact, to emphasise the 'archetypal nature of the scenes portrayed'.[78] Hämeen-Anttila, in turn (2002), has recently suggested that although the *Ḥikāya* is not a theatre script, it was probably inspired by the actions of party-crashers and jesters who would improvise entertainment in exchange for food.[79] In struggling to relate the dialogue in the *Ḥikāya* to a particular character or characters, these scholarly readings again clash over the palpably vexed relationship between realism and representation in the *Ḥikāya*, whose protagonist is at once archetypal and based on a real human being.

In his brief description and analysis of the *Ḥikāya* in his *Of Dishes and Discourse: Classical Arabic Literary Representations of Food* (1999), Geert Jan van Gelder calls Abū al-Qāsim 'not so much a realistic character as a protean mouthpiece of everything that undermines itself'.[80] Set within a broader context of the storied relationship of food and words in Arabic literature, van Gelder's reading also addresses the *Ḥikāya*'s possible contemporary reception, describing its obscenity as 'tolerated subversion'. 'It would be wrong', he goes on, 'to speak of a subversive text, unless one means by it that even the establishment likes to be subverted by way of titillation and as an occasional outlet.'[81]

Monica Balda-Tillier's character study of Abū al-Qāsim, 'Marginalité et éloquence contestatoire', further wrestles with the protagonist as a character (2003).[82] Abū al-Qāsim's often shocking performance, she says, as an imitation of the Baghdadi people, seems to suggest that 'to be a Baghdadi, is to be, at root, an *ʿayyār* (or brigand)', which is what she thinks Abū al-Qāsim himself is, his pious robes and religious recitations being a mere mask (an assertion with which I disagree, as detailed in Chapter 5, 'The Cosmic Crasher'; as the author himself describes his character in his conclusion, Abū al-Qāsim, a walking contradiction, is 'composed of [both] complete sincerity and complete hypocrisy').[83] Balda-Tillier's essay is especially useful in its comments on Abū al-Qāsim's relationship to language, noting that Abū al-Qāsim is *hors la loi* even of discourse (i.e. he does not converse according to the laws of normal human conversation), which is why his audience never manages to respond to him successfully.[84]

In introducing his work and its protagonist as a portrait of Baghdadis, the author has caused a great deal of confusion for readers of this text. What kind of portrait does it offer, and can it be said to be 'realistic'? In answering this question, scholars are often forced to contradict themselves, suggesting, for example, that the language is too unliterary to be realistic, or that its filthiness at the same times reflects an urban environment while also detracting from its depiction of an urban environment. Precisely this ability of the *Ḥikāya*'s language to, at once, obscure and reveal is central to my reading of the character Abū al-Qāsim as a trickster. I define the trickster as a character who engages in (apparently) improper, illegal, or immoral behaviours, but by doing so, stimulates the health of the greater community. According to my definition, the trickster is part of the divine while at home in the profane, and his main instrument in negotiating this tricky position is his language.[85] Like the language of the Qur'an, the language of the *Ḥikāya* is not wholly comprehensible, and it is this very obscurity or illegibility that provokes examination, commentary, and debate. It is disconnected from the 'real' in such a way that it suggests the possibility of other realities. It is at once a self-contained world and a reflection of other worlds, real and imaginary, in dialogue with past and foreign literary conversations, and provoking further debate and conversation today.

A Mediterranean Table

The association that early scholarship makes between the *Ḥikāya* and classical texts, and especially the *Satyrica*, survives and continues to influence later readings. Even in the introduction to his 1998 translation, René Khawam compares the two works, saying of al-Azdī, '… *On est en droit de le considerer comme le Pétrone arabe*' (We are entitled to consider him the Arab Petronius).[86] The *Satyrica*, now surviving only in fragments, undeniably has much in common with the *Ḥikāya*. Like the *Ḥikāya*, the *Satyrica* draws extensively from traditional literatures, and yet defies categorisation or definition with its unique structure that confirms to no one prior genre.[87] Like the *Ḥikāya*, its dating and authorship have been the subject of considerable debate,[88] and the history of its preservation and transmission is something of a mystery, especially given its obscene content. It has been called 'one of the most licentious and repulsive works in Roman literature', and yet (as Laurence Sterne described the transmission of his *Tristram Shandy*), it was able somehow 'to swim down the gutter of time'.[89]

Although this comparison has often been made, it has never been taken seriously as the point of departure for an in-depth study. Therefore, this study will compare the *Ḥikāya* to the *Satryica* and to other ancient and medieval banquet texts, and explore how these works play with notions of the representational qualities of literature. In doing so, it will examine how the *Ḥikāya* in particular works with or against literary tropes in its portrait of Baghdad and its inhabitants.

The introduction to the *Ḥikāya* quotes the assertion of the 'ancients' that man is a microcosm.[90] Medieval Arab thinkers' interest in the philosophy of the 'ancients' (by which they often meant the ancient Greeks) is well known, and similar descriptions of man-as-microcosm can be found across great gaps in time and cultural space, as can certain highly specific literary tropes having to do with banqueting, social parasitism, and more. Though it is impossible to account for these similarities exactly, the well-known Arab interest in ancient Greek medical texts could easily lead them into contact with more literary (as opposed to scientific) works, especially when it comes to texts about eating and food. Such subjects allow authors to do both science and literature at once (as in the *Deipnosophistae*, to take a Greek example, or *Daʿwat al-aṭibbāʾ* (*Physician's Dinner Party*) in Arabic, both of which texts will receive treatment in this study).[91] None will deny the Arabs' interest in Greek scientific and philosophical material, and the fact is that this material often blended seamlessly into what we now call 'fiction'.

Just as the strict categorisation of genres can limit our scope and understanding as readers, so can the perception of regions and cultures as hermetically sealed entities. As Tehmina Goskar writes, 'From a medieval Mediterranean perspective, the fragmentation of scholarship caused by modern disciplinary constraints – Islamic, Byzantine, Western – has artificially set apart cultures which had more in common than has hitherto been acknowledged.'[92] Baghdad and Isfahan, the cities featured in the *Ḥikāya*, were part of what the Greeks knew as the *oikoumene*, or the inhabited world, and this seems to me a more useful rubric to this study than the division of the ancient and medieval world into East and West. The idea of an inclusive Mediterranean literary tradition is especially attractive in the context of banquet literature, because the Mediterranean is round like a table around which many parties gather. Karla Mallette argues in her 'Boustrophedon: Towards a Literary Theory of the Mediterranean', that a major defining feature of 'Mediterranean literature' is an awareness of the existence of other languages and literary traditions and an (often playful) sense of competition between them. The *Ḥikāya* evinces this awareness not only through its comparison of two cities and cultures (including the relative beauties of the Arabic and Persian languages), but through its multifarious use of modes of speech, genres, and registers. Although it mainly depicts the speech of one over-talkative character, the *Ḥikāya* effectively presents a variety of voices gathering at a dinner table. This is typical of banquet texts.

In its presentation of many various types of discourse, the *Ḥikāya* resorts to various rhetorical strategies that stem from ancient traditions.[93] Consider, for example, the following passage, in which the alcoholic Abū al-Qāsim describes his favourite beverage:

> There is nothing in this world, by God, to equal wine as an antidote, shepherding the food to the depths of the belly, washing the body of corrupted humours, digesting the hidden puzzles of illness out of the stomach, food for the soul, a sister of the spirit ... By it the actions of nature are realised, digestion is improved, youths are made men, and old men are freshened. A balance, by God, for mankind in the four natures that resemble it: moisture is like its moisture, and its vigour and colour are like the vigour and colour of blood. The bubbles floating on it like foam are like yellow bile, and its dregs are like black bile. Every drink in the world depends upon it![94]

Many medical and philosophical ideas, such as the role of the four humours in the composition of the body, have found their way into the speech of the drunken party-crasher in this passage (and these connections are the subject of an article-in-progress, titled 'Medicine, Microcosm, and Mujūn'). Given the presence of these ideas within the multiplicity of voices with which the party-crasher speaks, and the preoccupation of the *Ḥikāya* with an all-inclusive microcosm, it seems most appropriate to read it in dialogue with other voices around the Mediterranean table.

The *Ḥikāya* is everywhere hailed as unique and innovative. In many ways it is unique, but elements of the text are easily traceable to trends in contemporary Arabic literature, such as an interest in jargon and the speech of outcasts, and in the character of the party-crasher in particular. Even more of its content and style may be illuminated if we do in fact expand our gaze to include a broader literary tradition of sympotic and banquet texts, or depictions of urban environments. For example, we find many similarities between far-flung representations of beggars with sharp tongues, or the marginally wise and mad, of which number we can count Abū al-Qāsim. Stories of the ancient Greek Cynics and especially stories of Diogenes, shown by Dimitri Gutas in his article 'Sayings by Diogenes Preserved in Arabic' to have survived in greater numbers in the Arabic language than in Latin,[95] may account for some of these similarities. As Daniel Kinney argues, 'the link between Cynics and professional parasite-jesters is as ancient as Plautus and Horace at least.'[96] Named after the dog, who both barks at people and fawns and begs, the shockingly outspoken Cynics, at once shameless and proud, self-sufficient and dependent on charity, often remind us of the similarly ambiguous Abū al-Qāsim. Their parodies of epic rhetorical styles and gnomic sayings twisted to 'justify transgressive eating'[97] are recognisable to any student of medieval Arabic parasites. Perhaps in parsing the character of Abū al-Qāsim, we may find evocative Branham's description of the Cynic Diogenes (whom Bakhtin calls a 'dialogic figure'): 'Diogenes' most brilliant invention was not a set of doctrines, let alone a method, but himself – a concrete yet malleable demonstration of a *modus dicendi*, a way of adapting verbally to (usually hostile) circumstances.'[98]

The transmission of these Cynic tales and other Greek texts through Arabic is well documented.[99] However, a full review of the network of cultural and literary currents responsible for the resemblances between ancient Greek, pre-modern Arabic, and other banquet texts and picaresque-style narratives cannot be provided in this short study, which instead focuses on a few evocative points of comparison, with both Greek and Roman literature (like the *Satyrica*).[100] These Roman literary traditions are bound to the Arabic traditions by shared culinary and hospitality practices, popular mystical beliefs, and reliance on ancient Greek medical and philosophical thought. Obviously the Roman Empire and later Arabic-speaking empires overlapped geographically, and there is a history of interaction between the regions involved, not only through trade, but through wandering entertainers and scholars. Furthermore the well-studied Latin literature (despite a similar prejudice in the scholarship against obscene texts), is equipped with volumes of commentary and theorising, much of which is informative when applied to the Arabic literary tradition.

Of particular interest to this study is the figure of the social parasite, and especially the party-crasher, in his many manifestations in Mediterranean literature. In a fragment of Petronius' *Satyrica* known as the *Cena Trimalchionis*, or 'Trimalchio's Dinner-Party', Encolpius, hero of the *Satyrica*, is not exactly a party-crasher, but is a similarly motivated outsider to the feast (as described in Chapter 3). Another outsider to ancient literary feasts is the Greek *parasitus*, a professional sponger who entertains with his witty conversation in exchange for food.[101] These social parasites share much in common with the Arabic *ṭufaylī*.[102] However, the *parasitus* is not a party-crasher because he is invited. But both the *parasitus* and party-crasher earn their bread in much the same fashion, with their eloquence and wit, and in some sense, the party-crasher is basically an uninvited *parasitus*. This study will refer to the party-crasher as a subspecies of the parasite, though many of the traits that I ascribe specifically to party-crashers can be found more generally among all parasites, such as their keen interest in food, often described with the same highly specific images. For example, they have unusually high tolerance for picking up and eating hot foods, which allows them to eat more quickly than their fellow guests, as found in Arabic and Greek sources alike.[103] Corbett describes the Roman *scurra*, a social parasite whose origins apparently lay in a kind of street theatre, and who would insult fellow dinner guests as a form of entertainment. These *scurrae*, with their penchant for mimicry and invective, also share much in common with Abū al-Qāsim (and are furthermore related to the Cynics).[104]

Amy Richlin's *Garden of Priapus* (1983) might suggest a fruitful field for further research on the *Ḥikāya*.[105] Richlin's study compares the aggressive stance of the satirist to that of the Roman garden god Priapus, who comically threatens garden thieves with rape with his gigantic phallus. Although Priapus is a lowly, often humorously self-deprecating god, he is nevertheless divine. Similarly, Abū al-Qāsim seems at once a lowly beggar and a cosmic symbol (as detailed in Chapter 5). Priapus is also a terrifying opponent not only of garden-thieves, but of witches and the evil magical power that they represent. Like the phallus itself, obscenity and grotesque and deformed features can be used to ward off the evil eye. Therefore, this Priapic position, at once comically abject and boastfully aggressive, paradoxically results in the self-aggrandizement of the satirist, even as he humiliates himself. In Roman and

Arabic culture alike, court jesters and dwarves could serve just this purpose at a banquet, where they simultaneously entertained and warded off bad luck with their unusual features.[106] This may explain why the often grotesque and obscene Abū al-Qāsim was considered an allowable addition to the Isfahani party that he crashed.

These comparative approaches can help us fruitfully situate a work within its broader contexts, a work which is itself strangely unique and self-contained. The *Ḥikāya* introduces itself as a microcosm, and my reading of it is influenced by my understanding of this text as a world isolated in itself, but containing a greater macrocosm. For epigrams for this study I have used quotations from *Moby-Dick*, a text similarly broad in the range of subjects it explores and registers of language it employs in doing so, and a text that is also meatily concerned with the 'real' while remaining elusive in its esoteric significances. The Pequod, an environment as isolated and self-sufficient as the party Abū al-Qāsim crashes, is usurped in its mission by the wild captain Ahab, a man as tyrannical as the Arab protagonist of the *Ḥikāya*. Like a ship at sea, the *Ḥikāya* owes its existence to the ingenuity and exchanges of various cultures, but its precise origins are a mystery, and its destination and the future reception of its strange and bountiful cargo is likewise unknown.

Informed by the nature of the *Ḥikāya* itself, this study is at once concerned with the highly specific and with the universal, and it is focused on the disruptive and creative potentials of language. Thus, I aim to provide a close reading using a deconstructive and comparative approach. The first chapter of this study, 'A Sampling of the *Ḥikāya*', provides some translations of key passages, which I will refer back to in subsequent chapters. The second chapter, 'A Microcosm Introduced', using the author's own introduction to his text, introduces some key themes: namely the tension between realism and mannerism, the literary representation of Baghdad as a greedy party-crasher, and the passage of time in the *Ḥikāya*. The third chapter, 'Crashing the Text', shows that Abū al-Qāsim is part of a literary tradition of party-crashing characters, and that in this tradition, the tension between the food and language serves to problematise the representational potentials of language. Chapter 4, '*Mujūn* is a Crazy Game', in focusing on a chess game between Abū al-Qāsim and another guest, explores the game-like qualities of *mujūn*,[107] as well some nonsensical and topsy-turvy language in the *Ḥikāya*. It also introduces some questions related to the status of this text and its protagonist as a microcosm, while discussing the microcosmic significance of the game of chess. Chapter 5, 'The Cosmic Crasher', delves deeper into Abū al-Qāsim's status as a microcosm, reading him as representing the paradoxical joining of the opposite figures of Satan and the Prophet Muḥammad. It shows that the trickster figure and the holy man often blend into one another in medieval Arabic mystical literature and beyond. The Conclusion explores the tension between the character Abū al-Qāsim as a cosmic symbol and as a weakening and mortal old man. It also compares the *Ḥikāya* to the story of Philemon and Baucis in Ovid's *Metamorphoses*, in which Zeus and Hermes, disguised as beggars, show up uninvited at Philemon's and Baucis' poverty-stricken home, which they subsequently transform into a temple of the gods.

Notes

1. I read the *Ḥikāya* as being in the iterative tense because it begins with the statement: *kāna min ʿādatihī an yadkhula dār* ('It was his custom to enter a house') *Ḥ*, p. 53. However, in a private communication to me, van Gelder said that the effect of this *kāna* ('he would') would have to last hundreds of pages in order for the story to be read as being entirely in the iterative tense. Without the effect of this *kāna,* the rest of the verbs, occurring plainly in the *muḍāriʿ* (imperfect tense), could express a non time-specific possibility, i.e. that he 'will' or 'may' perform an action. Consider also *Ḥ*, p. 332 (*yakūnu hādhā da'bahu* ('This is his habit[?]'). The fact that I cannot be certain even of the tense in which it is told is just the first hint of the general obscurity and confusion surrounding this narrative.
2. *Ḥ*, pp. 71, 106, 313 ff.
3. Charles Pellat characterises *Ḥikāyat Abī al-Qāsim* as the 'link in the chain' between the original use of the word *ḥikāya* to mean 'imitation or mimesis', and its modern meaning of 'story or relation' because it was written during a time the word seems to have changed in meaning ('Ḥikāya', p. 367). This is discussed in further detail below (p. 9) and in Chapter 2 (p. 71).
4. Translated by Bosworth, *Mediaeval Islamic Underworld*, p. 66. See de Goeje, 'Abulkasim', p. 723.
5. For example, 'Dieser Versuch, neben den ausgefahrenen Geleisen der alten Literatur einen neuen Weg zur lebendigen Gestaltung der Gegenwart zu bahnen, hat keine Nachfolge gefunden' ('This attempt to open up a new path

to the living form of the present from the tracks laid down by ancient literature, has never been emulated') (Brockelmann, 'alazdi', p. 1569). However, Jaakko Hämeen-Anttila, though he also describes the *Ḥikāya* as 'a work which falls outside the genres of Mediaeval literature', and which is '*sui generis*' (p. 84), asserts that it did influence the shadow play of Ibn Dāniyāl (*Maqāma*, p. 87).

6. St. Germain, *Placing an Anomalous Text*, p. 147.
7. Some of these are discussed in Chapter 3.
8. Al-Ḥusayn, 'Muqāḍāh', p. 17.
9. See Mez's introduction to al-Azdī, *Sittenbild*, pp. xiv–xv. However, see 'Baghdad the Party-Crasher' in Chapter 2 for mention of a poem cited in the *Ḥikāya* that was possibly composed later than this date.

   Hämeen-Anttila analyses two passages that occur in both the *Ḥikāya* and al-Hamadhānī's *Maqāmāt*, concluding that these passages fit more in their contexts in the *Ḥikāya* than in the *Maqāmāt*. This suggests that al-Azdī, therefore, probably did not borrow these passages from al-Hamadhānī, but that they both borrowed from an earlier source, 'possibly one connected with the circle of aṣ-Ṣāḥib ibn ʿAbbād, who was well known for his interest in the lower class in general and the Sasanians in particular.' By 'Sasanians', he means not members of the Sassanid dynasty, but beggars who (comically) claimed descent from this royal line. He also concludes that the probable date of the composition lay somewhere between 374 AH (a date early in al-Tawḥīdī's productive years), and 400 AH (984–1009 CE) ('al-Hamadhānī', pp. 83–96).
10. *Ḥ*, p. 390.
11. See Antoon, *Poetics of the Obscene*.
12. St. Germain's dissertation provides a complete consideration of this debate (*Placing an Anomalous Text*, pp. 10–35), accompanied by charts showing the birth and death dates of the names mentioned in the text, a review of the quoted passages from many contemporary texts and their probable dates of composition, and a discussion of the probable date of composition of the *Ḥikāya* based on this information. Like me, she concludes that debate on the authorship of the *Ḥikāya* cannot be regarded as conclusive, and therefore chooses to refer to the author by the name given in the introduction. She sets the possible date range of composition between AD 990 and 1020. Her analysis of these arguments is thorough in the extreme, and requires no further remark.
13. He began his career as a student of theology, but broke away from this field of study after visiting the Middle East (Bigger, 'Mez').
14. Al-Azdī, *Sittenbild*, p. xii.

15. For more on al-Jāḥiẓ, see Chapter 2, pp. 71–2 and 90, as well Chapter 5, pp. 138–48, 157, in which his depiction of man as a microcosm is discussed at length. For more on Ibn al-Ḥajjāj, see Chapter 2, pp. 74–5.
16. De Goeje, 'Abulkasim', p. 723.
17. Brockelmann, 'Alazdi', p. 1569.
18. Kraemer writes of al-Tawḥīdī that 'his humanism was not a joyful celebration of man's grandeur but a sober acceptance of man's ambiguity', capturing this author's undeniable affinity with the *Ḥikāya*'s protagonist, who embodies a portrait of mankind at once lovable and aggressively off-putting. Both Abū al-Qāsim and al-Tawḥīdī reflect, again in Kraemer's words, 'the entire spectrum from piety and asceticism to sacrilege and cynicism' (*Humanism*, p. 222).
19. These will be exhaustively detailed in Geert Jan van Gelder's and my new edition and translation (in progress, and mentioned below, p. 62n).
20. *Ḥ*, pp. 432–58.
21. See pp. 13–14.
22. See p. 62n.
23. For example, his chapter on 'Manners and Morals' refers to the *Ḥikāya* and to the works of Abū Nuwās (d. 814), who was known for his poetry about wine and the beautiful young boys who serve it. For more on Abū Nuwās see 'Those Camels Have Passed' in Chapter 2, p. 91.
24. Al-Azdī, *Sittenbild*, p. v.
25. Halliwell, *Aesthetics*, pp. 367–8.
26. Waardenburg, 'Mustashriḳūn'.
27. Macdonald, 'Ḥikāya', p. 304.
28. Al-Azdī, *Sittenbild*, p. xv.
29. Hämeen-Anttilla rejects attempts to relate the *maqāmāt* to ancient Greek literature, and especially to Greek mime. These attempts he sees as unfounded and unnecessary to explain the evolution of this type of Arabic literature (*Maqama*, p. 89). As for the translation of '*Ḥikāya*' as 'mimesis', Pellat writes that, although the Greek term was sometimes translated into Arabic this way, and although 'it is certainly possible that the idea of literary art as an 'imitation' of life might have produced the genre represented by Abū 'l-Muṭahhar', the Arabic tradition of mimicry mentioned in al-Azdī's introduction to his work seems sufficient to explain the *Ḥikāya*'s innovative project, and references to Greek literary influences, he implies, are not necessary (Pellat, 'Ḥikāya').
30. Horovitz, *Mimen*, p. 18.
31. Von Grunebaum, *Medieval Islam*, p. 289.

32. Von Grunebaum, *Medieval Islam*, p. 306
33. *Ḥ*, p. 44.
34. Al-Azdī, *Sittenbild*, p. xix.
35. Horovitz's article received a much-needed (though still brief) update by Hämeen-Anttila in the new third edition of the *Encylopaedia* (2010). Azarnoosh provides a lengthy and thoughtful overview of the entire text in the *Encyclopaedia Islamica* (2012), with a review of Mez's and Gabrieli's analyses of the text. He devotes special attention to its importance in the field of Iranian studies and literature, which is where his own academic interests lie. See my discussion of Kennedy's 'Maqāmāt as Nexus of Interests' below (p. 16) for a description of *adab* literature and its relationship to the *Ḥikāya.*
36. Mez writes that its language is 'too unliterary' to be helpful to the philologist (al-Azdī, *Sittenbild*, p. xx).
37. Gabrieli, 'Sulla *Ḥikāyat*', p. 34. Gabrieli identifies the physical description at the beginning of the *Ḥikāya*, which, he writes, differs from other more stereotypical descriptions found in *adab* literature, as one of the most important proofs of the realism of the character Abū al-Qāsim (p. 36). However, in comparing this physical description to other examples, Cooperson describes it as a 'manneristic elaboration', that 'does not seek to create a portrait of a particular individual' ('Images without Illustrations', p. 15). This again demonstrates the *Ḥikāya*'s ability to polarise opinions among readers in regards to its realism, with two scholars concluding that the description of Abū al-Qāsim is much more, and much less, realistic than other similar physical descriptions.
38. Gabrieli, 'Sulla *Ḥikāyat*', p. 36.
39. Lancioni, 'Gabrieli, Francesco'.
40. Gabrieli, 'Sulla *Ḥikāyat*', p. 36.
41. Schmeling refers to this work, frequently compared to the *Ḥikāya* (see below, pp. 11, 18), as the *Satyrica* in the interest of 'regularizing the title of this novel with those of other ancient works' (Schmeling, *Commentary*, p. xvii).
42. Davidson, 'Pleasure and Pedantry', p. 292. Similarly, Ceccarelli characterised Athenaeus as 'a sophisticated and mischievous author' who 'reminded [her of the author of] *Tristram Shandy*' (Athenaeus, A Walking Library?).
43. See pp. 89–91.
44. Al-Azdī, *Sittenbild*, p. xix.
45. This distinction was already blurred in medieval European and Arabic understandings of the term (Halliwell, *Aesthetics*, p. 341).
46. See pp. 15–16.

47. This thesis, which was originally in French, was published in a shortened version as *La Prose Arabe au IVe siècle de l'Hégire (Xe siècle)*, which briefly addresses the *Ḥikāya* (p. 130).
48. Mubārak, *Nathr*, p. 419.
49. Mubārak, *Nathr*, p. 419.
50. Mubārak, *Nathr*, p. 426.
51. Mubārak, *Nathr*, pp. 429–31.
52. Von Grunebaum, 'Aspects', pp. 259–81.
53. Von Grunebaum, 'Aspects', p. 281.
54. Von Grunebaum, 'Aspects', p. 275.
55. Al-Azdī, *Sittenbild*, p. xii.
56. Dhū al-Nūn Ṭāhā, 'Mujtama'', pp. 14–25. The definition of *mujūn* is discussed in Chapter 4.
57. Dhū al-Nūn Ṭāhā, 'Mujtama'', p. 16.
58. Ahsan, *Social Life*, p. 11.
59. Ahsan, *Social Life*, p. 162.
60. Sawād was a rural region in southern Iraq.
61. *Ḥ*, p. 168.
62. Goitein, *Mediterranean Society*, pp. 112–13 (with thanks to Nancy Spies, scholar of medieval and Mediterranean material cultures, for providing me with this quote).
63. *Ḥ*, pp. 135–7.
64. See the 'Editions' section above, p. 5, for more on this author.
65. Bosworth, *Mediaeval Islamic Underworld*, p. 65.
66. Bosworth, *Mediaeval Islamic Underworld*, p. 66.
67. Kilito, *Séances*, pp. 42–8.
68. Cf. Balda, 'Marginalité', described below in this chapter, p. 17.
69. Kennedy, '*Maqāmāt* as a Nexus of Interests', pp. 153–214.
70. Kennedy, '*Maqāmāt* as a Nexus of Interests', p. 164. Gabrieli similarly writes, 'The novelty of the *Ḥikāya* of Abū 'l-Muṭahhar in relationship to the *maḳāma* [...] is the displacement of the centre of interest from the purely linguistic and formal aspect to the representation of a character and an environment in a genuine *mimesis* of reality (in this case, the bourgeois environment of Bag͟hdād ...' (Gabrielie, 'al-Azdī', p. 31). Whether or not the *Ḥikāya* is best described as a 'genuine *mimesis* of reality' is a subject of debate throughout this present study.
71. Kennedy, '*Maqāmāt* as a Nexus of Interests', p. 153.

72. Kennedy, '*Maqāmāt* as a Nexus of Interests', p. 162.
73. Kennedy, '*Maqāmāt* as a Nexus of Interests', p. 167.
74. Kennedy, '*Maqāmāt* as a Nexus of Interests', p. 164.
75. See p. 122.
76. See p. 86.
77. Moreh, *Live Theatre*, pp. 94–103.
78. Rowson, 'Review of *Live Theatre*', pp. 466–8.
79. Hämeen-Anttila, *Maqāma*, p. 85.
80. Van Gelder, *Dishes*, p. 74.
81. Van Gelder, *Dishes*, p. 78.
82. Balda, 'Marginalité', pp. 371–93.
83. Balda, 'Marginalité', p. 383; *Ḥ*, p. 391. Balda further defends her position in ''*Udhrī* Love and *Mujūn*', saying that the positive characteristics of Abū al-Qāsim only serve to emphasise that he knows how to behave correctly, and that this knowledge is what allows him to affect a total reversal of the values of courtly love and polite behaviour (p. 129).
84. Balda, 'Marginalité', p. 392.
85. Hyde's *Trickster Makes this World*, discussed in the section 'Bojangles Won't Dance' in the conclusion (p. 170), was influential in the development of my own definitions of the trickster.
86. Khawam, *24 heures*, p. 10.
87. See Schmeling, *Commentary*, pp. xxx–xxxviii.
88. Schmeling, *Commentary*, pp. xiii–xvii.
89. See Schmeling, *Commentary*, p. xxi.
90. *Ḥ*, p. 44.
91. See Chapter 3, pp. 104, 112, 115.
92. Goskar, 'Material Worlds', pp. 189–204. James E. Montgomery similarly concludes in his 'Islamic Crosspollinations', 'We eradicate much … if we are too restrictive, too linear, too causal in our attitudes and interpretations for thus we often become deaf to the wider resonances of the phenomenon', p. 174.
93. The literary staging of a debate between two entities, for example, is an important feature of the *Ḥikāya*; see van Gelder's 'Conceit of pen and sword', for a discussion of the Greek origins of Arabic literary debate, pp. 330–5.
94. *Ḥ*, pp. 324–5.
95. Gutas, *The Cynics*, p. 480.
96. Kinney, 'Heirs of the Dog', p. 299.
97. Branham, 'Defacing the Currency', p. 93.

98. Branham, 'Defacing the Currency', p. 87. Branham usefully discusses the treatment of Diogenes's character in *The Dialogic Imagination* on the previous page (p. 86).
99. See also Gutas, 'Plato's *Symposion*'.
100. The *maqāma* genre, sometimes considered to include the *Ḥikāya*, may itself have influenced European picaresque traditions (see Monroe, *Picaresque Narrative*).
101. Sean Corner provides a very useful analysis of the Greek parasite or flatterer in his articles 'Politics of the Parasite' (Parts One and Two). Like me, he sees this character as inherently self contradictory; his main interest is in the contradiction between the parasite as a poverty-stricken beggar, dependent on charity, and the parasite as a gourmand, dependent on his own greed for fine foods.
102. Lucian's *De Parasito* is of special interest. For the Roman tradition of social parasitism in literature, see Corbett, *Scurra*; Anderson, *Second Sophistic*, pp. 183–5; and Damon, *Mask of the Parasite*.
103. *Ḥ*, 66. Davidson, *Courtesans and Fishcakes*, pp. 146–7.
104. Corbett, *Scurra*, pp. 41–2.
105. Richlin, *Garden of Priapus*.
106. In her famous but dated work, *The Fool*, Welsford in fact suggests that the medieval European tradition may have originated in 'the East', and in particular, as a result of Harūn al-Rashīd's emissaries to Charlemagne (p. 113).
107. Translated as 'licentious absurdity' above (pp. 13, 119), and further defined in Chapter 4.

1

A Sampling of the *Ḥikāya*

In order to provide the reader with an idea of the contents of the *Ḥikāya* and the varieties of registers and subjects that it includes, I will here summarise the events it portrays and translate some representative passages.[1] I will refer back to this section at various points in my argument.

Because the *Ḥikāya* is brim-full of quotations and doubtful readings, a fully footnoted version of the translation here, tracing all of its many quotations to their sources, would result in an unwieldy tangle of notes not directly related to my purpose, which is to provide the reader with the flavour of the text. As is repeatedly stated throughout this study, the language of the *Ḥikāya* is extremely obscure, and any attempt to translate it must necessarily be regarded as provisional.

The author first mentions some of the many sources from which he gathered his material, and introduces his text as an imitation of a Baghdadi man he once knew. The text begins with a traditional invocation of God and introduction of the author of the text. Classical Arabic works often begin with a praise of God, and the introduction here claims to abbreviate and replace this conventional section of praise of the Divine. It is likely, however, that such a section never existed, and that this itself is a mere literary convention adopted by the author.

> In the name of God, the Merciful, the Compassionate:
>
> The distinguished author, Abū al-Muṭahhar Muḥammad ibn Aḥmad al-Azdī, may God have mercy on him, after praising the Almighty as is His due, and wishing peace and blessings upon our lord Muḥammad and his family, wrote:
>
> The kinds of literature I am inclined to select are the speeches of the

Bedouins and old Arabic poetry, followed by the fantastic and flowery imaginings of well-read scholars, as well as the marvellous innovations born of the genius of the prominent modern poets.[2] These are the sources I have drawn from in my book, adorning myself with their work and often passing it off as my own; I have heard with my own ears the witticisms they have discussed at length and competed over. I have also included excerpts of my own poetry, letters that I have circulated, and records of literary gatherings that I have attended.

This is an imitation of a Baghdadi man whom I knew well for a time. He was always blurting pronouncements, sometimes pleasant and sometimes rude, as well as local sayings from his city, sometimes high-brow and sometimes shocking. I have preserved them in my mind to serve as a token of the manners of all the people of Baghdad, of all different social classes, and as a sample of their local customs.

It is as though this work has gathered them all together into one form under which their various types fall, each type of person sharing one definition in such a way that they do not differ, in spite of their various stations or their varying ranks.

Perhaps, I have thus become as Abū ʿUthmān al-Jāḥiẓ described in one of his books:

> Nevertheless, we sometimes see a man who can do impressions of the people of Yemen for example, recreating their unusual pronunciation in every particular. He could imitate a Moroccan, a Khurāsānī, a Persian, a Sindī, or an African, and yes, even to the point that he seemed more natural than they did.
>
> For instance, were he to imitate a man with a stutter, it would be as though the peculiarities of speech of every stutterer in the world were gathered together into one tongue. Likewise you might find him imitating the blind with a single expression that he would make with his face, his eyes, and his limbs, in such a way that you could not find, even among a thousand blind men, one who gathered all of these features together into one person. It is as though this imitator had gathered the features distributed among them all, and encompassed all of the elements of impressions of the blind in one blind man.

> There was a fellow who used to stand at the Karkh Gate where the muleteers gathered,[3] and he would bray in such a way that no donkey, including the sick, old, or overworked donkeys, could resist braying back. Had they heard a real donkey braying, they would not have paid him half the attention or been as stirred as they were by this donkey-impersonator. It was as though he had joined all the timbres of every donkey's bray into a single donkey bray, and all donkeys were moved in their souls to hear it. That is why the ancients claimed that a human being is a microcosm born of a macrocosm, because he forms every form with his hand, and imitates every sound with his mouth, and because he eats plants like beasts of burden, and eats meat like beasts of prey, and eats seeds like birds, and because in him are the shapes of all the kinds of animals.[4]

That said, I will add that this imitation amounts to the events of a single day, from its beginning to its end, or likewise, a night. One can just get through it and absorb it in that same span of time. For those readers who are eager and have the energy to hear this work out, and who do not find its long-windedness or literary overflowings to be a strain on their minds, and who do not consider the common expressions evidence of my faulty vocabulary or think less of me for using them, (especially when they arrive at the literary Bedouin imitation that I added to the end, and if they follow the saying that 'the spice of wit is low language, its charm is brevity, and its vitality in keeping it short'), for such readers I have gone to great pains, and have burdened myself so that they are left with what is enjoyable.

Furthermore there is precedence for my endeavour, as is expressed in a poem of Abū ʿAbd Allāh ibn al-Ḥajjāj which I will borrow and plunder here:[5]

> *Sir, this is a call to all those poets*
> *who write ordinary verse,*
> *please forgive these strange expressions of my*
> *playing foolish and perverse.*

A further precedence is found in his saying:

> *Take, good sir, my hand – the one I used to wipe my ass!*
> *I made a movement strange to see, a chessman standing fast,*[6]

like an egg that stands on edge inside a boiling pan.
Ibn Hārūn would be amazed, al-Bustī would have laughed,[7]
so come enjoy this oddity on which I've plied my hand!

He was right to say of himself, as I will say of myself after him:

All my speech is a late-night conversation, so sit back
and enjoy this late-night conversation of mine!

Now it is time for me to begin my tale, having already apologised for it, for as the poet remarked:

Usually I'm reticent and shy, but in good company, with noblemen I trust,
I give my temperament free reign, and speak my mind in manners most robust.[8]

The story itself then begins with a description of Abū al-Qāsim Aḥmad ibn ʿAlī al-Tamīmī al-Baghdādī, an old man with a white beard, red-faced, alcoholic, dirty, smelly, but also holy, a hermit, and above all, a big talker:

This man of many epithets,[9] Abū al-Qāsim Aḥmad ibn ʿAlī al-Tamīmī of Baghdad, was an old man with a white beard shining against the redness of his face, which almost dripped undiluted wine. He had eyes that were like they were looking through a dark glass,[10] gleaming like revolving balls in quick-silver. A punk and a low-life, a voluble loudmouth, a magic party-crasher, a wondrous litterateur, a clever reveller, praising and slandering, a gentleman-brute, an idiot-savant, close and distant, dignified and impetuous, buddy-buddy but two-faced, a late-night rambling gambler, a top and a bottom, roasting, zinging, berating, upbraiding, vituperating to your face or behind your back, cursing and reviling, a nasty drunk, a satirical saint, a heretic and a holy hermit, game for anything, a feather in the cap and a black eye,[11] an outrage, a disgrace, a hard-boiled world-weary pimp, a hustler, a scroll hidden in a cupboard hidden in a chest hidden in a tower, sealed with ambergris and wrapped in green silk ...[12]

Fouler than a fish-seller's slime, ranker than the fumes of a tannery. He grew up with Spike and Bruiser in bad company.[13] Always coming and going, a bundle of vice and a bucket of sin, a bag of mange and a robe of disaster, scrounged out of a garbage-can, a handful from the fire-stoker,

> a scrap on a manure-pile, more messed up than a pee-rag, older than the prophet's mantle,[14] dicier than mouldy cheese, nastier than a rat, son of a big-clitted old lady on a grey mule,[15] son of a widow who wove her cotton in the moonlight,[16] always crafting a disaster, a kink in the handcuffs, he mixed with gamblers and wine merchants, he took after transvestites[17] and monkey-trainers, he studied the fortune tellers' and the conjurer's science too.[18]
>
> *An old man singed by hell-fire long before he dies,*
> *you'll find him clever, quick, and in the art of outrage, wise.*
> *Jurisprudent, theologian, perceptive and reflective,*
> *Imam of dissipation and prophet-sent besides.*
>
> *If you're bent on blaming him (for it's his lot to be blamed),*
> *you'll strive to scorn the old man, and to make him feel ashamed,*
> *But you will find him stupid and as stubborn as an ass,*
> *called on by iniquity, unrepentant to the last.*

Then the narrative begins, describing how Abū al-Qāsim would walk into a high-end gathering in Isfahan, dressed in pious robes and spouting pious sentiments, moving himself to tears, though the other guests seem uncertain whether to smile or to weep:

> He would show up at the houses of the elite, cringing and pretending to be pious, like the most abstemious hermit, wearing a large cloak whose hood draped over his forehead, hiding half of his face. Whenever he saw a distinguished gathering of prominent people, he would show up whispering a recitation of the Qur'an, during the course of which he would stop and greet the crowd, then, in tender, emotional tones, present himself to the owner of the house, saying, 'May God bless your countenance with peace, and heap honour upon you'.
>
> Then he would take his seat, muttering the Qur'an for a long time, and then raise his voice somewhat to recite something from the word of God: 'Men who are not distracted by goods and trade from their remembrance of God, from their prayers, from giving charity, fear always the day in which all hearts and minds will turn, so that God may reward them according to the best of their deeds, and grant them increase of His favour, for He provides for whom He wills without reckoning.[19]

> The people would think that he had finished lecturing them, as he sighed so deeply it seemed his lungs might bleed, yet he would resume posturing and making a show of his piety, and continue until he saw that somebody in the crowd had started to smile. 'Oh hard of heart!' he would then say, with that same quiet, humble piety, after shedding tears and sighing deeply from his chest, 'How are you joyful thus when al-Ḥusayn was slaughtered?[20] There is no power or strength save in God! You are frittering your life away with amusements and diversions, though the family of your prophet suffered murder and war!'[21]

Finally some tough guy in the crowd intervenes, and at a critical moment, tells Abū al-Qāsim to 'come off it, because everybody here drinks and fucks'.[22] After hearing this, Abū al-Qāsim removes his pious robes and adopts a casual demeanour. One by one he then insults the other guests:

> 'My God, is it true what you say?', he would ask, 'All cuckolds? Dopes? Sons of fuckers and embracers on mattresses? Do they believe in meat and fried food? Do they worship a goblet or a glass of wine? Are they brothers of the fried sandwich? All of them? Yes!'
>
> He would loosen up his clothing and take off his hood, and he would sprawl out a bit, and say, 'Good morning! ... and no scandal!'
>
> Then he would look at one of the guests.
>
> 'Who, sir, is *that*?' he would say, turning to the host, 'What's his name? May God grant us the pleasure of his getting lost!'
>
> 'That is a man of exceeding erudite knowledge', the host would reply, for example. 'He is known as Mr Happy'.
>
> 'He frowned and turned away',[23] Abū al-Qāsim would say.
>
> 'There is no God but God!' he would say, '"Mr Happy!" That's like a total bore called "Mr Fun", or a shit-collector called "Fragrant", or a beggar-woman called "Her Majesty", or a sewer pipe called "Mr Clean!" It's like a black woman modestly veiled,[24] or a ruin with a padlocked door. I think that this one read *The Retardation of Knowledge*, or *The Book of Forgetting Stuff*.[25] He studied at the Institute for the Deficiency of Understanding. Someone probably picked him up at the cattle market last Wednesday. Nothing escapes him, praise God, when it comes to ignorance! Doesn't the old man not know that he doesn't know?

[. . .]

Who's that other one? He's like a picture drawn on a bathroom door!'[26]
'That is Secretary So-and-So', someone would say.
Abū al-Qāsim would say,

A 'secretary' to smack the truly literary person in the face with a sandal.

A 'secretary' who sits at the meeting,
farting silently in the noses of the great men of the age.

A 'secretary' who slaps 'Abd al-Ḥamīd in the back of the head with a sandal.[27]

A 'secretary' who smells shit and attacks like a soldier.

No, by God! Rather he's:

A 'secretary' whose doorman's shit is a better writer
than the beard of Abū Qurra.[28]

'This man is connected with the head of a government office!' someone would say, 'He is a very important man!'

'What's that to me?' he would say, 'The shit of a camel when the tide comes in! There hasn't been anyone worthy of awe since the prophet Muḥammad and his companions.

May the mother of this 'head of a governmental office'
by God's will fall a victim to a herd of horny horses!
He is, to me, a dog, or yet, a doggy's daily courses;
for there is no distinction [twixt dry faeces and its sources].[29]

Abū al-Qāsim goes on to insult the remaining guests, and when the host finally asks why he does not insult him as well, our hero replies that birds of a feather flock together, and that therefore the host and his guests are all a bad batch, not one of them gentlemen. Pressed by the crowd to explain how to be gentlemen, Abū al-Qāsim provides what would conventionally be seen as the opposite of good advice with regards to eating, sex, and financial debt. When a fellow guest laughs at this outrageous advice, Abū al-Qāsim reacts very negatively.

'Abū al-Qāsim', someone would say, 'What should we say? What should we do?'

'Be men!' he would say, 'Good, honourable men! Don't be beastly!'

'Abū al-Qāsim', someone would say, 'What should we do to be good men?'

'Live the life of the wise', he would say, 'Live by my code, and then that's how you'll be'.

'Abū al-Qāsim', they would say, 'Explain it to us'.

'There is no point', he would say, 'in telling signs and warnings to a people with no faith.[30] You can't make a dead man hear you. The deaf won't answer your call if they have their backs turned.[31]

'Were there life in you to hear me, you would hear me, but there's not.

'I'm selling pearls in the brick maker's market. It's like a pack of skittish donkeys, fleeing from a lion – deaf, dumb, blind, and no brain either.

What God gave me of intellect
is lost on donkeys, sheep, and cows.
They cannot hear me call, nor would
they understand me anyhow.
They gather up and croak like frogs
between the pond and willow boughs.'

'Abū al-Qāsim', someone would say, 'Get to the point'.

'But will you follow my advice?' he would ask.

'Yes', they would say.

'Do as I command you', he would say, 'and refrain from what I forbid you to do.[32] Receive my speech with obedience, for I live by the rules that I recommend to all.

'Those of you who are rich: don't store your wealth in fear of hard times, or save it for some no-good relative to inherit and squander.

'Those of you who are poor: Ask for a loan. Take on some debt. Don't mind the creditors and debt-collectors, but become connoisseurs of fine food, and drink liquor, and listen to beautiful singing-girls, and fuck the girls that don't sing, and the singers as well. Fuck when you're up and pray when you're down. Fuck the free-born, and don't forget the slaves. Fuck in secret and in full view. Fuck the owned and the free, and the whores and the chaste. Fuck as long as your cocks are standing, because they won't stand forever. Fuck the young and the old, fuck vaginas and assholes, fuck blooming young girls, and ancient old women, and bright young lads, and ugly old men.

True studs, they say, do not demur,
so climb on filth, and fuck a cur!

'Enjoy yourself with slave girls and slave boys, luxuriate in young women and men.

'Choose as your companion one whose reins are always loose, who with party-madness (*mujūn*)[33] causes the night to run on into morning, one who doesn't have a woman to go home to, no wife to tell him no or give him trouble, but rather a man who waves his cock from right to left, fucking those he mustn't and those he may. That is the clever artful sort, the noble young knight. Choose him as your dearest friend. Take him as a partner, as a brother. Join with him to fuck slave boys, the ones with small pricks and big assholes. Every fresh young slave boy in a girdled tunic – the ones who haven't changed and don't need to shave or use hair-removal cream yet.

Like the moon on a full-moon night, his beauty makes my chest grow tight!

'All of this advice, by God, comes from a man who wants the best for you.

This advice comes straight from one who's coming up behind
to march you to the King of Hell tomorrow, line by line.'

One of the people at the party would laugh.

'A cutting knife!' Abū al-Qāsim would exclaim, 'A slaughtering throat-plague, a shot, a stab, a citron thorn! It's a flirting fever, gall and vitriol, sawing through hard teak, an Ethiopian bubonic throat-plague! Did I say that God was Two?[34] Or Three?[35] Did I refute the Qur'an with poetry?[36] Did I break the tooth of the prophet of God?[37] Did I ransack his family's tomb? Did I fire a catapult at the Kaʿba,[38] or pelt it with menstrual rags? Did I defecate in the well of Zamzam? Did I hock the holy camel of Ṣāliḥ?[39] Did I speak of God as the Jews or Christians do?[40] Did I fornicate in the mosque of the Prophet, between his tomb and the pulpit? Did I shit on the Black Stone? Did I chop off the head of Ḥusayn, son of ʿAlī?[41] Did I cut off

> the hand of Jaʿfar ibn Abī Ṭālib?[42] Did I eat Ḥamza's liver?[43] Did I rend the flesh of one blessed by God?[44] So then, loser, *what* are you laughing at?![45]

The guests, growing tired of his offensive conversation, suggest that they talk about the weather in Isfahan instead. This leads to the beginning of a theme that will continue for most of the book: a comparison of Isfahan to Baghdad. Abū al-Qāsim switches between praising Baghdad and lambasting Isfahan, treating a number of features and amenities of the cities in the following order: overall atmosphere, place names, horses (at great length), cloth, houses, mats, perfumes, food (at very great length), the men who come to clear the food away, fruit, home decor, wine, snacks, drinking companions, and entertainers both male and female (at great length). Some of the lengthy descriptions of good and bad horses are presented here (a section so long it prompted complaints from Mez, as detailed above):[46]

> 'I have never in all of my time in Isfahan', he says, 'seen a good man on a good horse – a swift, spirited, high-stepping steed, quicker than the eye, drowning all description, stunningly formed. His breeding apparent, as though veiled with starlight, shoed in uncompromising stone, he vies with the falcons in their diving, outracing the arrow in its flight. He leaves the cheetah out of breath, towering like a mountain, or rushing in a violent torrent, a diving star or a blast of lightning, a flaming coal or a cloud downpouring. His reins are loose, his step is sure, and his chest is wide. As swift as the passing of a day, his body lent spare wings, he is a ship of the desert, the wind made flesh, the whip his reins, the spread of the earth his racetrack.
>
> *A son of the wind, his father the lightning!*[47]
> *[...]*
>
> *If released he melts like a wave, if halted he freezes,*
> *an antelope reaching with cheeks (what cheeks!) for food that it seizes.*
>
> 'Another:
>
> *As tractable as water, but when galloping, a fire.*
> *He spins just like a compass when you turn him in a gyre,*
> *you'd think he was a bird, were a stallion not his sire.*

'Imru' al-Qays said it well when he said:[48]

Charging, fleeing, front and back, a hurtling avalanche,
ostrich's leg, waist of gazelle, wolf's lope, and fox's dance.

'The new-fangled poets of our modern age said it well when they said:

Short twixt flank and back, long twixt neck and knees,
his ankles are Orion, his blaze the Pleiades.[49]

'He may be fair like a shooting star, grey like a ghost, black like a raven, bay like the sun behind a slip of clouds, yellow like melted gold, or dappled like a sword half sheathed.

Fair-haired, and in his face his fortune's ever rising,
his blaze is like it's silver melted onto gold.
His rider's like he's on a castle when he's riding,
night-racing like the magic horse the prophet rode.[50]
[...]

'But in Isfahan I only see a billy-goat riding a donkey, a bastard on a mule, a monkey on a nag. He backs up and flees, or kicks up and stumbles, unsightly, unruly, balky, and wild. He bites and he kicks ... When he gallops, he farts. The most notable of your nobles rides a swollen hack with a fat neck. Quite an uproar, coughing and farting together! A fart for every step, with a cough and a poot.

Kicking and surging, biting, bucking and twisting.

'He's bloated with straw like a sack of hay, a she-ass could outrun him, and he bolts at the squeak of a mouse. Or he might be skinny as the number 1, a deathly ill little leather bag, who stops at a tug and stumbles on a piece of dung. You could hobble him with a hair, his hide consumed with scabies, and a mangy tail and mane.

'With legs knobby as cucumbers, with calluses big as quinces and a hoof tumour like a watermelon ... he's more of an orchard than a horse!

Driving him and leading him is like pushing a ship with a punting pole.

'Another:

Dust coloured, one blue eye, one black,
His ears droop down, his visage falls.
He's bleary-eyed, so when he walks
He tends to knock against the walls.
A peck of ticks bedeck his mane,
And ant-like along his bridle crawls.

'Another:

He's blind and deaf, weak-backed, foot-sore,
with wobbly legs, hair scant, so old
he's outlived vultures by the score,
he leans on walls, he trips in holes.
He can't be forced on anymore,
A hand's-length's too far now, poor soul!'[51]

Having insulted the Isfahanis' mounts, he moves on to describe the inadequacy of their cloth compared to that which he implies is found in Baghdad:

> Nor do I see on any of your backs red-dyed Egyptian Dabīqī robes, or Dabqāwī, Qīrāṭī Zuhayrī, or Qushayrī gauze robes, or Aden cloaks, or Tākhtaj, or Rākhtaj, or linen clothes, or Tapestry, Dasīsī, Tinnīsī, Dimyātī, or Mujallalī either, or silk brocade, interwoven with gold, scented with ambergris, beautifully striped, as though woven from flowers of the springtime. No diaphanous Sīnīziyya gowns, like the very air in thinness, like a mirage. No serviettes of long silver stripes with which to pat the mouth at a party, nor feathered cloth, or sashes of Maghribī gold, or Dabīqī tabby lined with gold embroidery.[52]

For every description of Baghdadi luxuries, he insults the same things in Isfahan:

> I only see desolate houses on a squalid swamp, the walls fortified with mud and spattered with caca, spread with [local] Ruwaydashtī wool blankets, velvet cloths, and Kurdish haircloth, and Khāwarānī pillows. And you, summer and winter, sitting on mats of hair, calico on your bodies, coarse, Marvi, rough, home-spun, with holes for the farts to get out. Layered yarn: that's what your

> shirts are made from, and your turbans, which hang loose around your head and flop over your ears ... If you're dressing up, you go sleeveless, and your young slaves go in streaky cloth with bright blue cotton turbans with borders of green and red strips. If you wring out the shirt of your market-man you get a full jar of grease, with cheap perfume that smells like toothpaste.[53]

During his comparison of Baghdad to Isfahan, Abū al-Qāsim describes food at especially great length (and indeed such 'drooling speeches' are typical of party-crashing characters).[54] He begins with appetizers:

> Peeled walnuts, white and fresh, whose flavour with Dīnawarī or Byzantine cheese is sweeter than bodily health. Turnips, white and red, like lumps of fried dough, and the first suckling lambs, bringing light to the eyes, stimulating sexual desire, and cutting through bile, soaked in wine vinegar, imported from Ṣarīfīn and ʿUkbarā. Cucumbers with vinegar, asafoetida, pickled eggplants, and almond cake with pomegranate water.[55]

And he ends with desserts:

> ... *lawzīnaj*[56] stuffed in fine flat bread, flavoured with rose water and musk, with a thin crust and lots of stuffing, fried in almond oil, fragrant to smell, with honey like un-chewed gum, and caliphal *lawzīnaj* dry, musked, Abbasid, and fine *fālūdhaj*[57] with wheat kernels and the saliva of bees, and cool fresh water much intermixed with saffron and almond, with pearly grease, as if the almonds in it were stars shining in a sky of lightning. Barmakid butter paste containing fig syrup and bees' honey, and nice fried crepes drowning in julep, sitting in a crystal conical cup, stripped and sturdy, and colourful Chinese plates.[58]

Then he describes the servers who come to clear away a finished meal:

> Then the food is lifted away and in comes the waiter, radiant-faced, with clean clothes, perfect manners, light spirit, with a precious imperial toothpick in his hand, like a silver pole made by Najāḥ the Black, or a scented Maʾmūnī toothpick. He kindly offers it to the guests. This is followed with fine mahaleb cherries, scented and flavoured, from Sharika the perfume merchant. He bestows upon their hands, once rubbed with scent, white potash, mixed with ground cedar and Khurāsānī clay plus a little

> frankincense, chufa, sandalwood, ashed musk, powdered musk, camphor, and Bengal rose, imperial, royal, foaming like soap, frothing like the sea. The hands squeak with it, like an Indian shoe. It is from the shop of Ibn Ezra the Jew, who selects only the whitest potash, white like the shit of sparrows. He counts it one by one, then grinds it like powder, yes, and he brings a copper pan without peer, like the flame of an ember, or a lump of gold, and a silver pitcher of one piece, of the highest quality, with a handle joined to it with no break or seam, which, although light, holds a litre of water and is of marvellous workmanship. The crowd washes their hands with this, and then he offers them a Dabīqī napkin, velvety, light, embroidered, Egyptian, with two signet marks, two cross stripes, two dyes, finely made, the perfect length, the correct width, the fibres curled, bordered with a double border, embellished with raw silk, but softer than spun silk.[59]

Abū al-Qāsim goes on to describe the charms of Baghdadi slave girls (many of whom he mentions by name), and the ecstatic behaviour of their listeners (historical figures identifiable in contemporary biographical dictionaries and other sources). The section on slave girls begins with letters by Zād Mihr, slave of a certain Ibn Jumhūr, whom she openly despises. This Baghdadi slave girl, who 'knew nothing but the real world and hard cash',[60] shows no signs of submission, and indeed Ibn Jumhūr addresses her in his letters as 'mistress of her master'. Her bitingly witty letters in response represent one of the highlights of the work as a whole:

> This Zād Mihr, the slave girl of Abū ʿAlī ibn Jumhūr, was beautiful in the extreme, with a sweet voice, and the queen of the female companions. Her owner was one of the stupidest rudest people around, full of nagging, bickering, dalliance, and ennui. Once Abū al-Ḥasan al-Dawraqī visited him and demanded that she sing, so he wrote to her, during a time when she was somewhat peeved with him,
>
> 'Dear mistress of her master,
>
> 'Today I have a friend over, Abū al-Ḥasan, who came only to hear you. So I'd like you to favour me and come over, and no joking around, for the man is not a joker.'
>
> She wrote in response, 'He's the one I see with his moustache cut short! A shit-well, that's what he is. And I, by God, am unable to open my eyes

from this headache, and my throat is clogged up from the eggplant I ate yesterday!'

He wrote to her, 'By God, I have conveyed to him your apologies, but he isn't satisfied. He said, 'come today with the charitable donation of your song!'[61]

She wrote on the back of this note, 'Damn your eyes! And what if this our lord Abū al-Ḥasan (God bless him!) suggested further and demanded a lay, and said, 'Make it a charitable donation of your cunt this year!'? Tell him, if you please, that I swear to God I can't open my eyes! How many times do I have to say it, damn it, leave me alone! Rid me, O God rid me of you!'[62]

[...]

She also wrote to him,

'O Ibn Jumhūr, send me something to tide me over, and please me with clothing. Otherwise, by God, I'll go out and sing, and put my body up, and ten others with me. And you know that if a slave girl goes out a-singing, someone soon gets in her panties![63] I've warned you, and you know it full well! If you want all mankind to fuck me, I won't get in the way of your plans. I'll fulfil all your desires! You've got your own whores who suit you, and seven of them for a slap in the face. If you get up off of one of them, you get off of her with twenty farts in your sleeve. But still they brag about you, saying "we were with Abū ʿAlī the Sultan's merchant, the great and magnificent!" Yes, it suits you, the likes of that stupid female donkey in your house! You can crack a walnut on her head and she won't dare say a word to you, because she thinks you're minister Ibn al-Zayyāt or Ibrāhīm ibn al-Mudabbir. As for Zād Mihr, who pounds you like bulgar wheat and grinds you down like flax, well she's not something from your spice rack! By God, this house of yours in Basra is nothing so much like a madhouse, and I am one of the madmen locked inside! May the Lord spare me from my sins as he spared me from the sight of you, for I have become the happiest of people due to your long absence! But from suffering these financial tribulations I am wearing out my body and losing my youth, waiting on you. And all the while you forget about me, fooling around with your loser buddies in Baghdad, while I am in Basra sitting on reed mats and rags!

'Damn you Ibn Jumhūr, burn your eyes! You've become a sodomite, friend of slave boys with peach fuzz![64] God protect me from your wantonness! When the weaver's belly's full, he thinks his daughter a princess! By your life, I'm going out to sing and get fucked in Basra, while your boys in Baghdad rent out their wares, and you can be in the middle, Ibn Jumhūr, with a happy spirit. I'm not going to judge your actions, even if you're sometimes friends with boys, and sometimes women! By the life of your crooked nose, your eye-liner and your hair-do, I can compete with you, blow by blow! If you get into boys, I'll take lovers, if you get into girls, I'll do some tribbing. But I'll do you one better, because you're never wanted unless you pay gold, while I am desired and paid gold for doing it! And in the ass of the one who comes up short, a stick! ... I expected nothing less from someone like you, that you would lose interest in me. Well I have lost interest in you, so if you fall in love, I will fall in love with someone better than you, and if you get married, I will marry someone more elegant than you. Damn you, it's like you're always flying off the handle. You forgot me, got distracted from me! Send your dear mistress some money, and bring her to you from Wasit, before she starts to get angry. And by my life, give me a lute to use with a teak border and ivory inlay ... so that I can go singing with it. Fie on you Ibn Jumhūr! How quickly your forgot what you used to say to me: "No sleep can satisfy me until I hold it in my hand, then I fall asleep." Or perhaps you found one greater than it, softer, hotter, and tighter, and that is what has distracted you? Damn you! By my life, tell me the truth about it, even if the truth is something alien to you!'

This is just a drop from the flood of what she wrote.[65]

After describing the genius and charm of many such Baghdadi slave girls and other singers, Abū al-Qāsim lapses into nostalgic memories of his home town, of relaxing in gardens and extemporising poetry with his friends.

At the end of the day we slept between sweet-smelling plants, fanned by the breath of the orchard, when Abū ʿAbd Allāh,[66] drunk and blurry-eyed with sleep, suddenly noticed a line of boats passing on their way to Baghdad, and improvised this poem to them:

O ships of Baghdad, Go! You know,
My heart goes today too.

O Baghdad ships, what'd be the harm
were I one of your crew?
My breathy sighs will be the wind
that drives you, day and night,
and if you drift to shallows,
my tears lift you upright.
O ships! – Call of a lover
yearning for the loved-one's sight.

O ships of Baghdad, speak for me
To those for whom I ache.
I am a stranger in this land,
The birds sob for my sake.

Then sleep overcame him, but when he woke up in the middle of the night, and heard the moaning of a pigeon on a branch, his heart ached, and he groaned, and recited this:

Turtle-doves, stir my heart with your cooing
And wake me from this lengthy sleep,
And help me in sorrow and mourn me in death,
and speak for one grown far too weak,
Speak to the wind and tell it, 'O wind
of the north, pause and grant me relief,
and blow through my heart, for by this you will heal
With the air all the wounds of my grief.
O wind of the south, perhaps you can serve
As my messenger far down the road
To the wakers in Yaḥyā Souk, there where they stayed
In Ibn Ḥajjāj's abode,[67]
To those too who left me, my heart melting, here,
And weakened my body so worn,
For long have they kept me from sleep, and have said,
Floods of tears must prevent his return. [68]

Then he asks the host of the party for food. He is asked what he would like to eat. After responding with elaborately literary demands for fine food and expensive presents, he eventually simplifies his request, and is given some of the light snacks (cheese and pickled foods) that he requested:

> Then he would turn to the owner of the house and say, 'Now you've given us a headache! Bring us our breakfast! Verily we have found fatigue in this our journey'.[69]
>
> 'OK', the host would say, 'What do you suggest, Abū al-Qāsim? We've grown frightened of your disapproval.'
>
> 'Not at all!', he would reply, 'I don't wish to inconvenience your meal, God forbid!'
>
> 'Speak, Abū al-Qāsim', the host would say.
>
> And he would say,[70]
>
> *I want from you a loaf of bread, all on a table clean.*
> *I want some coarse-ground salt, I want some vinegar that's keen.*
> *I want some well-cooked meat, I want some vegetables plucked fresh.*
> *I want a suckling kid, or else I want a new lamb's flesh.*
> *I want some water cooled with snow in vessels clear and brittle,*
> *I want some presents, nor will I be satisfied with little:*
> *I either want a racing horse who, flying, races for me,*
> *or a line of singing girls to gather and adore me.*
> *I want a little gazelle fawn, all grace in every part,*
> *and like the moon, and soft and kind, and easy on the heart.*
> *I want a big fat rump, I want a gratifying cock.*
> *I want a shirt, a robe, and then a turban on the top.*
> *How nice that I'm your guest today, and that you are my host!*
> *So nice how you oblige me, as I don't wish to impose!'*
>
> And they would say, 'O Abū al-Qāsim, is that all that you want? It's a tall order, by God! No, you have to cut it back.'
>
> And he would say,
>
> *My heart yearns after rice and milk, and I'd like a good stew.*
> *If you mentioned fried meat, I could eat a dish or two.*
> *Rice pudding makes me happy when it comes so soft and white.*
> *Zīrbāj*[71] *is always loved by all whose head is screwed on tight.*
> *Our host has all this in his house, on this I know I'm right!*
> *[...]*

> 'Abū al-Qāsim!' someone would say, 'You're only making us more nervous with these overtures!'
>
> 'God forbid!' he'd say.
>
> 'So speak', they would say.
>
> 'Damn you!' he'd say, 'A nice soft loaf, some cheese so strong you weep, strips of local meat, tender and congenial, something from the ready foods of the market, and whatever lingering little bites you have around, like pickled snacks. Why are you making it so complicated? What's with all this fussy presentation?'[72]

He eats and then insults their food. Then he persuades another guest to play him in a game of chess. He roundly insults his opponent while playing the game, discussed at length in Chapter 4, '*Mujūn* is a Crazy Game' and translated here in full:

> Then he would wash his hands and say, 'Where are Abū al-Jalab and Abū al-Ṣannāj?' (by which he meant the games of chess and backgammon). And they would bring out, for example, a chess set.
>
> 'Who's up for it?' he would say. 'What poor sap wants to offer up his head?'
>
> They would be reluctant to play. 'Yes, when the governor arrives', he would say, '*Ruqayqīm* hides!'[73] until one would finally agree. When he noticed him he would say, 'The pharmacist meets his medicine![74] Now isn't this Mr Terrible about to become Mr Terrified!'
>
> 'So how does the poor bastard play?' he would then ask.
>
> 'He's a good player!' they would say.
>
> 'Well an old mule isn't frightened by the tinkle of a bell',[75] he would reply, and turn to him, saying:
>
> *You who challenged me, you've thrown a fire on dry sticks!*[76]
> *Now when this boxthorn sticks you, you're going to get a prick!*
>
> And he would begin by advancing his pawns, and reciting some nonsense by way of opening the game:
>
> *We went out early, at dawn, in the night, in the evening, after midday,*
> *And hunted rabbits, jackals, and wolves, but donkeys got away!*

Then his fellow player would advance some pawns, and he would say, 'Hey bum, don't bite off more than you can chew! Just two squares at a time, so you don't end in the black![77] Camel by camel or you'll break the *maḥāmil*.[78] I say enough, but he sneaks on up. Your basket won't split mister! Don't hurry, my lord, hurrying's for tomcats. He takes two of my pawns for one pawn – now that's a good deal!

Every time he sold his beard, I sold a hairless asshole.

An elegant man, by God!

I gave it to him good, and he found it rather thrilling.[79]
'Your beard up in my ass', I said, and he was more than willing.

He would advance the Queen protected by a pawn,[80] and say 'Go up with a blanket and come down with a fan'.

He would limit his opponent's play from the sides, and say, 'In Mrs Curves's crack, for it's made of solid rock!

A gift from me sent into you, in myrtle, basil wrapped,
A peach on bottom, topped with knob of apple, pomegranate.[81]

But if his opponent broke out and gave him trouble, he would say:

He slept, but with a shitty shoe,
I slapped him so he'd wake up,
Now look, the veins show in his neck
Just how they liked the shake-up.

His opponent would send his knight into the centre after the advancing of the pawns.

'Well done!' Abū al-Qāsim would say, 'Now we've moved on from playing jacks to spinning tops!' And he would say, 'The morning found me occupied with what I did all night, thus one keeps at a thing until mastering it. So, good sir, shit and play with it, that way you can do two things at the same time! Stand on the river bank and tie up what washes up.'[82]

'You could move the king's guard, dumbass', he would suggest. 'Your bread is covered in fish paste.[83] If you weren't plotting over something you wouldn't be eating your bread in the corner.'

And if his opponent would take up one of his pawns in his hand, and act as if he were going to move it, he would say, 'If you see the chicken pecking the rooster's ass, you know that she's telling him, 'fuck, fuck!'

Then his opponent would stammer, his error apparent to him, and Abū al-Qāsim would say, 'So OK, the blind man shits on top of the roof and thinks that no one can see him.[84] You dumbass! The one who farted in your beard ate beans from my farm! Your hand is closer to heaven than to that piece, and one who strives to stomp on the wind is but farting in his moustache.'

Then he would say to one of the people there, 'Why don't you watch this game and see something really amazing?' This would garner a little interest among some of the crowd, who would begin cautioning and advising his opponent, much to the annoyance of Abū al-Qāsim.

'Good sir, I told you to watch', he would say, 'I didn't tell you to get in the way! Leave him alone so he can get his finger jammed in the door, and then I'll show you how I slap him!'

But his opponent would be distracted. 'Damn you, what do you expect?' he would ask, 'For piping always distracts the piper from eating flour dry!'[85]

If he would groan out loud worrying over something, Abū al-Qāsim would say, 'He sings the song of a hornet in his clothes! He took a break from his work to cry about his mother-in-law. How he raves, God bless him, like a Hindi divorcee!'[86]

If someone says to him, 'Take that pawn in exchange for one of your pawns', and he saw that there wasn't any benefit in doing so, he would leave it and say, 'If it's a monkey for a monkey, better to take the monkey that's house-broken.'

Then he would take one of the pawns on the side of the board and say,

If you can't get a rose, take a cyclamen.

'Many a thing you despise at first turns out to be worthless.'

His opponent would then take one of his pawns. 'Too bad Abū al-Qāsim!', someone would say, 'Why did you sacrifice it without compensation?'

'Go to hell', he would reply, 'And bring a bit of firewood!'[87]

Then he would take the queen or a knight on the opposite side.[88]

'Good sir', he would say, 'a blow of the stone hammer is better than a thousand taps of the mallet.'

'Never mind', his opponent would say.

'If you hear someone say 'never mind' in a war, you know that the shit's above his head', he would reply.

His opponent would neglect to strengthen his position, and then realise it and begin to catch up.

'After the fart, he tightened his ass',[89] Abū al-Qāsim would say.

And his opponent would want to swerve to the side with his knight, but then would see that this was not possible for him, and Abū al-Qāsim would say, 'You dumbass! If they let you go on the hajj, take the straight path!'[90]

So he would return it to its place.

'The seed rolls round and round, then returns to the mill',[91] Abū al-Qāsim would say.

Then his opponent would softly mutter over something evincing worry, concern and vexation, and Abū al-Qāsim would say,

O you whose anger has led him to press
On the dregs of my ass with his fangs!

How long will you grieve? How long will you gripe? How long will you grow angry?' Then he would say, 'Poor thing. What's he doing? His flour fell in the thorn bushes and he can't get it back together again.'

He would play for something and his opponent would thwart him, and he would scream, 'Hey! He stopped me, by God, by sword and by flame! What do I do?'

But then his opponent would make a mistake in his play, and surrender some pieces to him, and he would say, 'Nice going, boneless doggy bag![92] Put the spoon of your face in my ass!'

Then his opponent would take a piece, and see his error and want to put it back, but Abū al-Qāsim would force him to take it and say, 'By God you'll take it even against your will!'

'And what should he do with it?' someone would say.

'What the slave girl of al-Sukkarī did', he would reply.

'What did she do?' they would ask.

'Take it in her hand and put it in her cunt', he would reply.

Then he would turn to him and recite,

Your cheek turns after the evening meal
To the black-hairy hole of my ass.
So be content with the bitter truth,
Or otherwise, just let it pass.
And if you are angry, tomorrow
do like al-Sukkarī's lass.

Then he would say, 'This, by God, has been the way of the game since its invention, until it bore this fruit that it bore. Yes, for the donkey is on hire until death.[93] Free straw bursts the sacks.'

Then his opponent would tempt him to take a piece, and he would stretch out his hand to take it, thinking it was free. But the error of his thinking would appear to him, and he would neigh and scream and recite,

O son of whom my pickle jar penis
comes and goes in her ass's avenues!

Another:

O you who hurry towards destruction
visiting this place,
the rook of my hand is coming down
to slap the king of your face.

'Dumbass! If you jump on two pegs, one of them is going up your butt.'[94]

Then his opponent would turn to someone as if to ask advice. 'If the sea-turtle has need of a boat, he's finished', Abū al-Qāsim would say.

But the person would advise him to do something, so Abū al-Qāsim would turn to him and say, 'Sure, take it from someone whose intellect fits in a fruit-basket. You grew so long, my cock, that you came out of my sleeve!' And he would recite:

What dire time is this that they go trampling on my game?
At my expense I'll teach this stupid ass.
I don't play favourites to the beards a-rubbing in my bum,
Except my friend here, him I'll get to last.

By whom he meant the host of the party. Then someone would say to him, 'Damn you, you can ask advice too if you like, just stop insulting people!'

And he would say, 'May the back be broken of the mother of the one who has to eat beans in order to fart!'

His opponent would strike at both his king and his bishop,[95] and Abū al-Qāsim would scream and say, 'I'll give you some advice, sir, by God!'

'What advice?' he would ask.

And Abū al-Qāsim would reply:

Tie up your beard, for deep in an ocean of shit
you have now fallen down.
A sea from my anus which looks like it's drawn
With a compass, so perfectly round.

Then he would say in a sing-song voice:

Father of Ḥusayn, son of al-Ḥasan,
Your head and your body grow wide!
Your beard too has grown and I wish that it would
Deep in mi culo *go hide.*[96]

Then raving he would say,

Umm Razīn shat in the flour one day. 'Why so?',
We asked. She said, 'It is the leaven in the dough.'[97]

'What does this cuckold care about?' he would ask, 'A hard head and firm horn!' and he would recite: . . .[98]

Another:

O youth whose soft black beard hangs lank
Like long loose strands of silk

. . .[99]

It's wrapped up in your mother's shit
and sealed up with a fart.

Another:

O vilest, lowest of mankind, in my view, without doubt,
I copied cutting verses with my shoe between his eyes.
His head's allowed for plundering, ears, shoulders blown about,
By power of my hands, now try my shoe-slap on for size.
Fear God, the cartilage of your ears, the weak veins of your snout![100]

Suddenly the king and rook would lie exposed to Abū al-Qāsim's bishop, an admirable elegant move, and his opponent would leap up in surprise. Abū al-Qāsim would talk nonsense in a sing-song voice, saying, 'This is odd manners, good sir! This is the uncouth language of Baghdad's Bāb al-Ṭāq,[101] and the strange whims of chance!'

Then he would say, 'A chess player on his death bed, as he was giving up the ghost, counselled his son thus: "Beware, my son, the side of the rook! Fear the pounce of the knight! Heed the leap of the bishop! For it is better to sit on the cock of a donkey than to sit in a square open to attack!" And then he died.

'Sound advice, by God! A religious duty! He gave his son his due and left him his legacy, may God not bless his carcass nor water his grave!'

This play would end with his opponent's head spinning. 'Alas for you!' Abū al-Qāsim would say, 'This young man, God bless him, came to my party today, but do you know what he ate?'

'No', they would say.

'A thousand cocks in a loaf of bread!'

His opponent would respond with a curse, which he would tolerate. 'Poor you!' he would say, 'For the loser is allowed to mock and scorn, while the winner should be tolerant and kind. So I don't blame him, by God. At this knot the carpenter farted.'[102]

His opponent's king would be in a tight spot, and he would say, 'Too bad! You escaped into a corner.' And he would recite mockingly:

He said 'spin it', I said 'in her cunt
Would that it would, good sir, spin . . .'

Should his opponent's pawns be scattered, he would try to gather them together with crafty manoeuvring, holding them back, and Abū al-Qāsim would say, 'When the shepherd dies, the sheep scatter'.

'And has the shepherd died, Abū al-Qāsim?' someone would ask.

'Half of him died', he would say, 'and the other half is on its way.'

'How's he really doing, good sir?' someone would ask.

'He's in the shit up to his throat', Abū al-Qāsim would reply, 'And the dogs are standing guard. He's doing about as well as beetroot in a hot pot. He shat in the pan, or rather on the chessboard. He shat in his own coffin. His juices are dry, and only the dregs remain.'

His opponent would rush to evade him with one of his pawns, and someone would say, 'How quickly he moves his pawn!' To which Abū al-Qāsim would reply, 'Someone said to a tent-peg, 'How quickly you go in!' and it answered, 'If you knew what was beating my behind, you'd understand!'[103] He moved faster than a half-inserted penis or a fleeing gazelle with his young running before him', he would mockingly remark.

His opponent would be stumped, and Abū al-Qāsim would say, 'Night-blindness in the day-time is blindness all the time'.

And he would say, 'Good sir, we slapped that monkey until we went blind', and he would recite,

Incense-like your shit did fall,
Oh what a transgression!

'Abū al-Qāsim', someone would say . . .[104]

'Yes, there's nothing for it but a draw! As the poet once said,

The time comes for a truce when truce becomes the only way.
Not peace but poop and in your beard is what you get, I say!'[105]

He would be planning to checkmate his opponent with his knight, and then he would strike it, saying, 'All right! Take a white thing for your coal, you scum!' And he would knock the chessboard over on his face.

Someone who was absent at the time of the victory would ask what had gone on between them. 'We were slapping one another', he said, 'and now he's complaining of a weak constitution!'

Then he would look around at everybody and ask, 'Are we fasting today?'[106]

Abū al-Qāsim then demands more food, and the narrator mentions a number of dishes that might hypothetically be brought to him and what his response to each dish would be:

> They would bring out a *sikbāj*,[107] for example, and 'This, by God', he would say, 'is the most trusted resting place of the stomach.'
>
> Finding it sour, he would say, 'Yes sir! The sourness of this vinegar causes the forehead to perspire and the stuffy nose to bleed. By God it's sourer than a cold-morning snow slap on a freshly-shaven head!'

He has a poem or joke ready for every kind of food mentioned. He ends this section by describing a great cook he once knew in Baghdad, an Ethiopian named Nāranj:

> He would satisfy in a moment, understand the slightest hint, anticipate any desire, as if he knew one's innermost thoughts, both of visitor and host. By God, he could cook something to help if you were sleepy, or bereaved, or drunk, or worried, and when he had completed a certain dish, and someone said, 'Nāranj, what do you need?' He would say, 'Just some hungry people!'

Then Abū al-Qāsim demands and drinks a glass of Isfahani water. At this moment he changes his tune about Isfahan, which he now praises to the skies, having formally characterised it as a filthy and cultureless city. Rather abruptly he begins a story about two anonymous men eating in Baghdad. One of these men he enthusiastically praises while the other he drags through the mud, all in more or less generic terms. This passage, described further in the section 'Baghdad the Party-Crasher' of Chapter 2, illustrates his willingness to praise one thing and blame another at random (as he does with the cities of Baghdad and Isfahan). He continues describing his food and drink, while praising Isfahan.

A guest asks why he changed his tone so abruptly, having formerly only insulted Isfahan and praised Baghdad. Abū al-Qāsim shrugs the question off, and goes on to describe in vivid terms the discomforts and drawbacks of living in Baghdad:

> Haven't you ever heard this about Baghdad?:
>
> *Let not Baghdad kindle your hopes.*
> *Its hope is deceit, and its nearness is distant.*
>
> May God refresh the bones of Ibn Muʿtazz for saying:[108]

How can I sleep now I've stopped in Baghdad
and will stay in her lands evermore?
In a country in whom all the wells wear crowns
of hovering black-flies galore?
Whose air in spring, winter, and summer is smoke
and whose waters as black as smoke pour?

And he would say:

Long was my time of worry in Baghdad,
whose visitor must grieve or leave that place.
I lingered there unwillingly, much like
a flaccid man in a dowager's embrace.[109]

Since he is describing Baghdad, his audience presses him for more insider information. Abū al-Qāsim responds with a long list of swim strokes and sailor's slang which, as he does not define them, are difficult to understand:

> During this conversation, someone would say, 'Hey Abū al-Qāsim, do you know anything about swimming?'
>
> 'Idiot!' he would reply, 'Does a country boy know how to ride a cow? Does a Turk know how to shoot a bow? I, by God, am swimmier than a frog or a sea monster. I know how to swim in ways that fish and ducks don't know anything about. I know the crawl, the arm-stroke, the underwater stroke, the back stroke . . .[110]

The other guests press him to describe his house in Baghdad, and he is offended by their prying, but describes it as a den of iniquity:

> Then one of them would ask, 'Abū al-Qāsim, where do you live in Baghdad?'
>
> . . .
>
> What business of yours is my house, damn you? Idiocy and meddling! A house built not on piety, praise God![111]
>
> . . .
>
> A house that has written on the door:
>
> *Whosoever enters this house is safe from all but fucking.*

Another:

Every night it shelters as its guests
a whore, a pig, wine, and gambling.[112]

He continues to praise Isfahan, the wine they are drinking, and the host of the party, who again inquires why he seems to have forgotten about Baghdad, which he formerly lauded. Abū al-Qāsim rails against his home town yet again, and begins praising a guest on his right, and then insulting him to his neighbour on the left, again demonstrating the overall fickleness with which he chooses to insult or praise something:

> Then he would turn, in this middle of these discussions, to the one on his right side, and he would strike up a conversation, listening to his talk and appearing well pleased with it. 'Sir', he would say, 'this is not the speech of a man! It is magic that drives the heart and the ears mad! Speech, by God, like a cool glass of wine and the mantle of youth. Or rather like Paradise on earth, and blooming youth, cut flowers and a sorcerer's knot. It's like nothing so much as the announcement of the birth of a noble son to a sterile old man, the beauty of brocade, the clarity of glass, the sweetness of swallowing that which cures the ill and restores the ailing. It leads his listener to prostration in prayer, and runs with the flow of sap through wood. Praise God, the way of verbosity is open wide to him, and the path of prolixity gapes open upon him, for he strings pearl after pearl.
>
> Then the person on his left would say, 'What are you up to?'
>
> He would wink at him with his eye, and turn to him and say, 'My lord, I am in the tribulations of the bald-headed woman who cannot grow hair, in a conversation heavier than stone and more bitter than colocynth, the delirium of the feverous, and the darkness of the sorrowful. This kind of thing makes the mute glad that he cannot speak, and the deaf man glad of his deafness. Speech, by God, that rusts the mind if it doesn't strike you blind. Speech that troubles hearing with its ruggedness and baffles the imagination with its defectiveness. It cannot reach the hearing nor be received in the natural mind.'[113]

He then praises the musicians at the party, growing more and more drunk. He suggests that the guests all continue to drink the next morning, essentially inviting himself to spend the night. Growing even more intoxicated, he flirts with a pretty singing girl present. He then verbally attacks her guardian with particular venom. Another guest laughs, and as at the beginning of the story, Abū al-Qāsim turns on this laughter with fury:

> Are you laughing at me, you son of a farting humbug, who farts wet sharts and sells them for the cost of an acorn, may God blacken your face you son of a farty filly in heat, shouting, blind, torn, foaming, asking to be fucked … bursting, braying, wood-pecking, wet, may God make my ass a flint bucket and your beard a bit of tinder?! A small-pocked monkey-mongerer in Baghdad by the low wall of al-Khuld is hooded with the thighs of your wife, and his cock is in her belly to the limit of the pith, you son of a dejected wino! If your mother's labia were Hāshimī trimmed with a forelock, I would have plucked its moustache in Medina mosque, inside the [holy] Maqṣūra. By the life of her mottled ass and the wet hair of her pussy, the stork-like kernel of her clit and the black-and-white magpie hair of her ass, I will pluck your tattered moustache![114]
>
> *O son of a big-clitted woman whose ass has turned aged and silly.*
> *The cocks play in her hole and her shit, shuffleboard, willy-nilly.*

The Isfahani guests finally grow sick of Abū al-Qāsim, and resolve to rid themselves of his conversation by encouraging him to drink until he passes out. As he drinks, he again insults them one by one. He then flirts with the singing girl, as well as an attractive boy in the gathering, and sings and dances until he falls down. He pesters the singer for more song, and the singer, irritated, calls him a plague in Persian.[115] He understands the insult, and with hurt pride, responds with a lengthy boast and further derision of his fellow guests:

> You dog, you went digging and found a privy, you prodded a sewer pipe with your cane! Direct your eyes towards me! Lend me your ears! Do not move your hands or your shoulders, and call up the stragglers! […] Damn you, do you know me or not? I eat sand and shit out stones! I swallow date pits and shit out palm trees! Damn you, I am the roiled wave, the complex lock, the fire, the brigand, and the grindstone turning! I walked two weeks with my head cut off! I invented villainy, and defined brigandry!

I am Pharaoh! I am Haman! I am Nimrod Ibn Kanʿān! I am Satan uncircumcised! I am the bare-fisted bear, the stubborn mule, the cruel war, the angry camel, the lusty elephant! I am Fate overthrowing, I am hardship oppressing! I am the tyrannical lion! I am the trumpet of war, the drum of strife! I am God's power parting the red sea! I am the Divine Decree! I am warning! I am stone! I tear up the ranks! I cut down two armies! I am known on every horizon for the chopping of heads! I am springtime in a drought! I am the rich man when bankruptcy appears! I'm more famous than the Eid of Ramadan! I am sharper than iron! I am the anchor! [...] I am one-eyed Willy! If the devil saw me he would run away! I am the Artful Dodger! I am the plucker of stone bridges! I am more rightly-guided than the sand grouse, more alert than a magpie, more persistent than house-flies, more dogged than a dung-beetle, sharper than limestone and arsenic, dearer than an antidote, more bitter than colocynth, and more prominent than a giraffe. When held up in the jungle I lived by eating beasts of prey and made the grass my vegetables! My food is bitter roots and blood is my drink and viper brains my candy! I cut my veins with every dagger and crushed my bones with every hot-rock. Always in prisons or dungeons, I cut creature's livers with patience. I saw a ghoul giving birth and carried Satan's bier, broke the jawbone of the tiger and saddled a lion. I have killed a thousand and I seek a thousand more! I'm on my way to the afterlife. I've accepted bribes. Do you need anything from Mālik, Treasurer of Hell? Damn you, do you know me? [...]

By God's light! An egg from me is worth a thousand; if it were hatched a thousand demons would come out. By the Lord's power! If my neck were cut I would not die for a year. If a man talked to me whose head was higher than Capella and whose feet were playing hopscotch [on the earth],[116] by the Lord's power, I'd just strike him a blow that would scatter his bones so that they couldn't be gathered in a month, or I'd pierce his nose and put it on his head, and slap him with it. I'd dislodge his head with two pounds of his shit! If a man talked to me whose head was made of iron and whose body was made of copper and whose legs were made of lead, I'd give him a slap that would make his nose fly off the back of his head. If a man talked to me who extinguished fires with his moustache, I would tie his nose hair to his pit hair and turn him around until he smelled the gas passing from his

> ass-gate. If I snorted, the monasteries of the Christians would fall prostrate, and the palaces of the children of Israel would crumble! I am Zurayq the Demon. Pharaoh could not scowl in my face, or stand by my side, or match me word for word in debate![117] My head is an anvil, my beard is a dagger, my moustache is a handlebar,[118] my canine tooth is a butcher's knife, and my hand is an ironworker's hammer! Perhaps someone has something to say, O son of the Bitch-slapped, O son of the Man-chaser? Perhaps you have a word to say, O son of the Fast-one, the Spread-eagled, the Scandal? Bark, you dog! Fill your eye with me! Fill it with a devil named Saqlāb who will use you like a shuttlecock ... If I didn't fear for the earth's safety I would snort a snort half thunderbolt and half earthquake. Damn you, by God, I will put you in my pocket and forget about you until you go rotten! I'll cut off your head and make it a button for my shirt! I will snort you up and only sneeze you out in Hell! I will drink you up and only pee you out on the Path of Righteousness! And Adam yelled, 'Oh what a loss!' I will gulp you down then fart you out, then return you to everything that harms you. Damn you, do you know me?[119]

Finally, with one last loving description of the beautiful boy who has taken his fancy, he passes out.

The next day he wakes up and resumes the pious manner in which he entered the gathering, reciting the same religious verses as at the beginning of the text. He scolds the guests for their celebrations when they should be occupied with more pious thoughts, and leaves them with a *salām ʿalaykum* (peace be upon you). The narrator closes by describing him once again as a man of extreme opposites and a representation of the people of Baghdad. These closing passages of the story are discussed in the section 'The Microcosm is a Man' in Chapter 5, 'The Cosmic Crasher', and also in the conclusion.

Notes

1. I am currently working with van Gelder on a full translation and edition of the *Ḥikāya*, and will save exhaustive footnoting for that volume, titled *Ḥikāyat Abī al-Qāsim al-Baghdādī al-Tamīmi*. Although the translations excerpted here are essentially my own, van Gelder's contributions have been such that it is no longer possible to distinguish my work from his. A few of the phrases were translated by him outright, and all of them were influenced by his instruction.

I, of course, take full responsibility for any errors, and for my decision to err on the side of looseness in favour of readability.

2. By modern poets he means urban poets from the middle of the second/eighth century onwards.
3. A market in the south-western outskirts of medieval Baghdad.
4. *Al-Bayān wa-l-tabyīn* (Book of Clarity and Clear Expression) I: 69–70. See 'Introduction', n. 3 (pp. 71–2, 90, 138–48, 157), for more on this author.
5. St. Germain describes the relation of these and other citations of Ibn al-Ḥajjāj in *Ḥ* to his *Diwān* in *Placing an Anomalous Text*, pp. 18–19. See the 'Authorship and Dating' section in the introduction for more on Ibn al-Ḥajjāj (p. 4).
6. Movement here means both of the bowels and on the chess board, and is a translation of *dast*, which signifies both 'stool/evacuation' in Persian, and 'game/place/trick' in Arabic.
7. Abū al-Fatḥ al-Bustī (d. 400/1009) and Bishr ibn Hārūn were both poets.
8. *Ḥ*, pp. 42–5.
9. From *muḥallā*, which we connect with *ḥilya*, 'epithet, description of a person'. Al-Shāljī reads *mujallī* (a horse first in a race). Shmuel Moreh writes, 'The senseless sentence in which he is characterised as a *maḥallī*, "a local person", is clearly the result of one of the many copying errors in the which the book abounds. *Maḥallī* should be emended to *maḥkī* or *muḥākī*: "this impersonated person/impersonator was known as Abū 'l-Qāsim … al Baghdādī"' (*Live Theatre*, p. 97). St. Germain translates as 'exiled' from *jalā*, 'to expel' (*Placing an Anomalous Text*, p. 156).
10. Literally 'green glass', but the colour green was often used to signify darker colours shading into black.
11. Literally, 'the white blaze on a horse's face', and 'scabies' or 'mange'.
12. *Ḥ*, pp. 46–9.
13. 'Spike' and 'Bruiser' are loose translations for several strange, foreign-sounding names that might have been nicknames for members of the criminal underworld or beggar class.
14. The prophet's mantle, which was later destroyed by the Mongols, would probably have been almost four hundred years old in al-Azdī's day.
15. For more on the insult 'big-clitted', (*baẓrāʾ*), see Batten, 'Making Men and Women'.
16. Cotton was thought to deteriorate in the moonlight.
17. Or effeminates. See Everett Rowson, 'The Effeminates of Early Medina'.

18. The word here translated as 'fortune-tellers' refers especially to false practitioners of astrology (see Bosworth, *Islamic Underworld* II: 257).
19. Q 24:37–8.
20. Ḥusayn ibn ʿAlī, grandson of Muḥammad, was killed at the battle of Karbala (680 CE) during the power struggle that followed soon after the prophet's death. His death is mourned during the Islamic month of Muḥarram.
21. *Ḥ*, pp. 53–4.
22. *Ḥ*, p. 55.
23. Q 80:21.
24. St. Germain (p. 161) explains that veils were meant to conceal beauty, 'but it was generally assumed that black women were not beautiful', and directs us to the following source for more information about this form of racial prejudice: Bernard Lewis, *Race and Color in Islam* (New York: Harper and Row, 1971), pp. 11–15.
25. *Ta'khīr al-maʿrifa* (translated as *The Retardation of Knowledge*) was actually written by the famous ninth-century wit and writer of nonsense, Abū al-ʿAnbas al-Ṣaymarī, see Ibn al-Nadīm, *Fihrist*, ed. Flügel, p. 152.
26. That is, the door of a bath house. The precise meaning of this insult is unclear to me.
27. ʿAbd al-Ḥamīd (d. 750), secretary in the Umayyad court, was a seminal writer of Arabic prose.
28. Abū Qurrat al-Ḥusayn ibn Muḥammad al-Qunnāʾī al-Kātib was an influential financial administrator in the late tenth-century Būyid court, known for his beautiful writing. See *Ḥ*, pp. 58–9, n. 2.
29. *Ḥ*, pp. 56–9.
30. Q 10:101.
31. Q 80:27, 52:30.
32. Enjoining good and forbidding evil is one of the duties of a Muslim, and is mentioned multiple times in the Qur'an (Q 3:104, 3:110, 7:157, 9:71).
33. The definition of this word is discussed in Chapter 4, p. 119.
34. This refers to the Manicheans, who proscribed to a dualist view of the divine (divided between light and dark, or good and evil).
35. This refers to the Christians (Q 5:73), who believed in a holy trinity.
36. This passage so occurs (with variations) in al-Tawḥīdī's *Akhlāq al-wazīrayn*, pp. 493–4, and will be further discussed in an article-in-progress on the relation between the heretic and the party-crasher, co-authored with John Turner, as mentioned in the Acknowledgements, p. 7.

37. A man was said to have broken the tooth of the prophet Muḥammad, who then forgave him. See St. Germain, *Placing and Anomalous Text*, p. 193, n. 485.
38. Al-Ḥajjāj (r. 738–44) fired a catapult at the Kaʿba when attacking Ibn Zubayr.
39. Q 7:73–7.
40. Q 5:74, 9:30.
41. Ḥusayn was the son of the prophet Muḥammad's cousin, ʿAlī ibn Abī Ṭālib (discussed in Chapter 5, pp. 151–2). He was decapitated in 680 after the battle of Karbala, an event still mourned by Shīʿī Muslims today.
42. He was the standard-bearer in the battle of Mu'ta against Byzantine Christians. They chopped off his hands.
43. Hind bint ʿUtba ate the prophet Muḥammad's uncle's liver after the battle of Uḥud (625 CE).
44. Abū Luʾluʾa, a Christian slave, stabbed ʿUmar ibn al-Khaṭṭāb.
45. *Ḥ*, pp. 82–6.
46. See p. 11.
47. *Ḥ*, p. 115.
48. Imruʾ al-Qays was the most famous pre-Islamic Arabic poet. St. Germain locates this poem in his *dīwān* in *Placing an Anomalous Text*, p. 212, n. 585.
49. This poem is by Ibn Durayd (d. 933), a famous poet and philologist. Although he had passed on decades before the probable time of composition of the *Ḥikāya*, he was modern relative to the poet of the previous verse, Imruʾ al-Qays. See al-Ṣāwī, *Sharḥ Maqṣūrat Ibn Durayd*, p. 69. Thanks to van Gelder for this reference.
50. *Ḥ*, pp. 119–120. Al-Burāq was a steed who carried Muḥammad to Jerusalem and to heaven during his famous 'night journey'.
51. *Ḥ*, pp. 127–9.
52. *Ḥ*, pp. 133–4. Many of these types of cloth are described in Serjeant, *Islamic Textiles*. Material cultural readings of the *Ḥikāya* are readily available to those interested in these subjects, as detailed in the 'Scholarship' section of the introduction, pp. 13–14.
53. Or rather, gum used to clean the teeth. *Ḥ*, pp. 149–50.
54. Gowers, *Loaded Table*, p. 62.
55. *Ḥ*, pp. 152–3.
56. Almond baklava. For a recipe see Perry, *Baghdad Cookery Book*, pp. 99–100.
57. Almond sweetmeats. For a recipe see Perry, *Baghdad Cookery Book*, p. 100.

58. *Ḥ*, pp. 162–4.
59. *Ḥ*, pp. 164–6.
60. *Ḥ*, p. 232.
61. *Zakat ghinā'ik*; *Ḥ*, p. 230.
62. *Ḥ*, pp. 229–30.
63. *Sarāwīl*, *Ḥ*, pp. 476.
64. The growth of the beard marked adolescent boys as being too old to attract the sexual attention of grown men. A similar theme is found the literature of ancient Greece.
65. *Ḥ*, pp. 236–8.
66. Ibn al-Ḥajjāj, a poet who lived in Baghdad, on whom see the 'Authorship and Dating' section in the introduction (p. 4).
67. See previous note.
68. *Ḥ*, pp. 270–1.
69. Q 18:62; translated by Pickthall. The Qur'an is an especially difficult work to translate. In this study I sometimes use my own translations, and sometimes use those of two of its most famous translators (Marmaduke Pickthall and Abdullah Yusuf Ali) according to whichever best serves to illustrate a given point. This same verse is given as the only Qur'anic verse that the famous party-crasher Bunān bothered to memorise (al-Khaṭīb al-Baghdādī, *al-Taṭfīl*, story no. 166).
70. Cf. al-Hamadhānī, 'al-Maqāmat al-sāsāniyya', in *Maqāmāt*, p. 202, in which this poem is included but missing the line about the gratifying cock.
71. A stew made of meat, vinegar, almonds, and spices. For a recipe see Perry, *Baghdad Cookery Book*, p. 33.
72. *Ḥ*, pp. 274–7.
73. I have been unable to identify the meaning of this proverb. Mez renders this *raqīquhum* (their slaves (al-Azdī, *Sittenbild*, p. 93).
74. Specifically *bizr qaṭūnā* which, according to al-Shāljī, are black seeds with multiple medicinal benefits, including the lowering of fevers, still in use in Baghdad today (*Ḥ*, p. 279). St. Germain, who identifies them as 'flea-bane seeds', adds the creation of erections to their medicinal utilities (*Placing an Anomalous Text*, p. 327).
75. A proverb. Mez refers us to Freytag (*Proverbia* I: 207). Also see al-Maydānī, *Majmaʿ al-amthāl* I: 167.
76. Specifically, *ʿarfaj*, a flammable plant.
77. This translation is based on al-Shāljī's assertion that *tā* was a Baghdadi col-

loquial shortening of *ḥattā*, now pronounced *dā* (*Ḥ*, 280). See also Dozy, *Supplément* (دا).

78. Litters for camel-back, which presumably could be become entangled and break if the camels walked side-by-side.
79. The meaning of this line is unclear. A more literal translation might read, 'I moved him in love, and when the move got power over him, he coughed' (*nāqaltuhu fī al-hawā munāqalatan fa-hiya idhā qaddarat ʿalayhi saʿala.*)
80. See Wieber, *Schachspiel*, pp. 314, 322.
81. Clearly an extended metaphor for a penis. Cf. Abū al-Qāsim's earlier taunt: 'That's right, honey, you want something planted at one end, with a nozzle at the other. It's not an eggplant, and it's not a gourd … I think you want something that starts with a truffle, ends with a cucumber, and has a nose-bag hanging on its neck. You'd like a good blow on the proverbial horn!' (*Ḥ*, p. 78).
82. 'What washes up' is an educated guess for *mābāqāt*, used earlier in a similar context: 'It's as though he's the Grand High Treasurer of Chicken-shit, or the Trustee of the River Bank in charge of all the duck droppings, or the commissioner of the Tigris tying up the *mābāqāt* with palm fronds' (*Ḥ*, p. 61). Also see Steingass, *Dictionary*, p. 1136: '*mā bāqī*: the remainder, the rest … *mā baqīya*: remaining over, rest, remainder, remnant; arrears, balance, surplus.'
83. See *bunn* in Dozy, *Supplément*, p. 116. He provides a recipe for this fermented fish paste condiment.
84. A proverb. St. Germain (*Anomalous*, p. 328) refers us to Freytag *Proverbia* II: 169. Also see al-Maydānī, *Majmaʿ* II: 65.
85. Meaning unclear. For an alternate reading of this line see Mez, who reads the piper with his pipe (*al-zāmir bi-zamrihi*) as al-Zāmir ibn Murra, Ibn Murra being a nickname for the devil (al-Azdī, *Sittenbild*, p. li).
86. Al-Shāljī recognises this saying as one popular in tenth/eleventh-century Baghdad (*Ḥ*, 282).
87. Literally 'a *dānaq*'s worth of halfa grass' – *dānaq* being a small coin, a sixth of a dirham (a standard silver coin).
88. Because the queen could only move one diagonal square at a time (Murray, *History of Chess*, 225), its loss would not be quite the devastating blow that it is in the modern game.
89. A proverb meaning something like 'to shut the barn door after the horse has bolted', which according to al-Shāljī is still used in Baghdad today with a slight variation ('after he farted, he fell silent' (*ṣamat*) (*Ḥ*, p. 283)).

90. Literally, 'the road of Ctesiphon' or al-Madā'in, a city south of Baghdad and consequently between Baghdad and Mecca.
91. A proverb. St. Germain (*Anomalous*, p. 331) refers us to Freytag, *Proverbia* I: 419. Also see al-Maydānī, *Majmaʿ* I: 297.
92. The word for doggy-bag (*zalla*) also means mistake, and this is perhaps some kind of pun.
93. Al-Shāljī informs us that this saying is still in use today in Baghdad in much the same form (*Ḥ*, 286). See al-Maydānī, *Majmaʿ* I: 297.
94. This proverb is also found in al-Tawḥīdī, *Baṣā'ir*, ed. Wadād al-Qāḍī; Maydānī, *Majmaʿ* II: 387; and al-Ābī, *Nathr al-durr* VI: 51. Thanks to van Gelder for these references and many others.
95. The bishop (*fīl* or elephant) could move only two squares diagonally (Murray, *History of Chess*, p. 225).
96. The Arabic poem includes the phrase 'in my ass' in Persian, *ba-kūn-i man*, here translated in Spanish, which currently in the United States has a similarly competitive relationship with the predominant spoken language (English).
97. This fuller translation (as compared with that used in Chapter 4's discussion of nonsense, p. 125) is the result of van Gelder's input, and thus one example of how our joint translation and edition will display improvements over the sample translations used in this study.
98. The meaning of the poem here omitted is too unclear to present a translation, and the beginning of the second line appears corrupt in the manuscript. A tentative translation would look something like this: 'O husband of her who sold her ass for an i.o.u., | return my penis's semen with a lentil dish on top. | Don't you see my shoe is curly-toed and Daybulī? [see *Ḥ*, p. 288 and al-Azdī, *Sittenbild*, p. li] | With it I come to you and stuff your narrow collar.'
99. Part of this line is missing in the manuscript.
100. The last two lines of this translation indulge in poetic license. The second to last line in the Arabic plays on the similarity between the sound of the words 'hand' and 'power', and 'sandal' and 'light' ('Under the hands (*aydin*) by whose powers (*aydun*) they freely administer sandals (*khifāf*) to the head not lightly (*khifāf*).' In the original poem 'snout' in the last line is actually 'neck'.
101. A neighbourhood in Baghdad (Le Strange, *Baghdad*, p. 218).
102. Mez adds only that 'this must be a saying', (al-Azdī, *Sittenbild*, p. li). He does not, however, suggest a possible interpretation. Van Gelder found this saying in the section *amthāl al-ʿāmma* in al-Gharnāṭī, *Ḥadā'iq al-azāhir*, p. 344, in non-standard Arabic: *fi-ākhir* [sic] *ʿaqd yidrāṭ* [sic] *il-najjār*.

103. For this saying Mez (al-Azdī, *Sittenbild*, p. lii) refers us to Socin's *Arabische Sprichwörter* (p. 203), where we find it translated: *Man fragte den Pflock: 'Warum gehst du in die Mauer hinein?' Er antwortete: 'Weil Jemand, der hinter mir ist, Gewalt anwendet'.* (Someone asked the stake, 'Why do you go into the wall?' It answered, 'Because someone who is behind me uses force').
104. It would seem there is a lacuna here in the text.
105. The original poem uses a play on the words 'truce' and 'shit', which sound similar in Arabic. According to Mez this is an old joke found also, for example, in the poetry of Ibn al-Ḥajjāj (al-Azdī, *Sittenbild*, p. lii).
106. *Ḥ*, pp. 279–92.
107. A sweet and sour meat and vinegar stew, with vegetables, nuts, raisins, and figs. For a recipe see Perry, *Baghdad Cookery Book*, pp. 30–1.
108. Ibn al-Muʿtazz (d. 908), who preferred the city of Samarra (then home to the new royal palace) to Baghdad, was the son of the Abbasid caliph al-Muʿtazz and, during a time of political turmoil, himself the caliph for one day, until he was murdered. He was far more successful as a poet. St. Germain locates these poems in his *dīwān* in *Placing an Anomalous Text*, pp. 351–2, n. 1204–5.
109. *Ḥ*, pp. 308–9.
110. *Ḥ*, p. 313. The swimming strokes grow increasingly difficult to translate after this.
111. An inversion of Q 9:109.
112. *Ḥ*, pp. 322–3.
113. *Ḥ*, p. 329.
114. *Ḥ*, pp. 349–50.
115. *Ḥ*, p. 370.
116. *Dabbūq* is defined in various dictionaries as a well-known game played by youths. The rules of the game are not described.
117. *Ḥ*, p. 375.
118. See al-Shāljī, p. 377, n. 2. Nāfarūt is a moustache that stands up.
119. *Ḥ*, pp. 374–8.

2

A Microcosm Introduced

But no more of this blubbering now, we are going a-whaling . . .

Melville, *Moby-Dick*, 'The Carpet-Bag'

Ḥikāyat Abī al-Qāsim is a text enveloped in mystery. It exists, as I have already mentioned, in only one undated manuscript attributed to an otherwise unknown author and identifiable in no contemporary sources. The contents of the text are no less mysterious than the envelope, filled with difficult language, often nearly incomprehensible. For the moment, some of these unknown words and certain questions about the *Ḥikāya*'s history and authorship must go largely unexplained. Some of the puzzles of its unique narrative style, however, are very gracefully explained in the introduction to the text, written by the mysterious author himself, who seems aware that his bizarre literary offering is bound to raise questions. And although he states his project quite plainly, the project itself is somewhat complex, so we should first attempt to unravel the author's own explanation of his text, thus introducing most of the problems addressed in this study. A full translation of this introduction is found at the beginning of Chapter 1, above.[1]

One of the notable features of this text is its lack of chains of transmission (lists of sources cited in an anecdote), despite the fact that in many ways the *Ḥikāya* is a collection of quotations. But the entire work is attributed to al-Azdī, as if by a third person, at the beginning of the cursory *ḥamdala* (praise of God), constituting a kind of *isnād* (chain of transmission) for the entire text to follow. In his own voice, al-Azdī then begins his introduction by mentioning his sources, including poetry old and new, Bedouin speeches, and all of his personal experiences as an attendant of literary gatherings and a man of letters. Having thus briefly explained the ocean of quotations to follow, the author states, '*Hādhihi ḥikāyatun ʿan rajulin baghdādiyyin, kuntu uʿāshiruhu burhatan min al-dahri*' (This is a *ḥikāya* of a Baghdadi man, whom I knew well for a time).[2] In his article on the word *ḥikāya* in the *Encyclopaedia of Islam*, Pellat states that *ḥakā ʿan* can mean, in a chain of transmission, to relate word for word on the authority of someone.[3] Some works of classical Arabic prose often contained collections of short anecdotes (*akhbār*, plural of *khabar*), each preceded by a chain of transmission (*isnād*) that tells the original source of the story and who subsequently told it to the compiler of the present text; we could read al-Azdī's use of the phrase *ḥikāya ʿan* as an attribution of the entire text to the 'Baghdadi man he knew well for a time', which source is represented by the character Abū al-Qāsim (although it is not suggested that Abū al-Qāsim was the man's real name).[4] Pellat identifies *Ḥikāyat Abī al-Qāsim* as the 'link in the chain' between the original use of the word *ḥikāya* to mean 'imitation or mimesis', and its modern meaning of 'story or relation'.[5] Thus, the elastic title written on the manuscript, *Ḥikāyat Abī al-Qāsim*, could plausibly be translated as *The Mimetic Performance of Abū al-Qāsim*, either in the sense that Abū al-Qāsim is the performer or is the object of imitation, or *The Imitative Relation on the Authority of Abū al-Qāsim*, in which case Abū al-Qāsim, or the friend of al-Azdī that he represents, would be the authority for a lengthy imitative quotation of the people of Baghdad, or, finally, it could simply mean *The Story of Abū al-Qāsim*. Given al-Azdī's treatment of the word in his introduction, however, this last seems less likely.[6]

So, having stated that this is a story, mimesis, or relation on the authority of an unnamed Baghdadi man, the author goes on to describe, not his acquaintance, but the contradictory speech acts of his acquaintance, which, he says, he 'preserved in [his] mind as a token of the manners of all the people of Baghdad'. He has strung all the people, high and low, together under the same 'paradigm' (*ṣūra*)[7] in such a way that he became, he says, himself a *ḥākī* (mime) of the sort al-Jāḥiẓ describes.

Al-Azdī then quotes a passage from al-Jāḥiẓ's *Al-Bayān wa-l-tabyīn* (Clarity and Clear Expression),[8] occupying more than a third of the introduction, in which al-Jāḥiẓ describes the sort of *ḥākī* to whom al-Azdī compares himself. This *ḥākī* seems first and foremost to imitate the accents or dialectical peculiarities of different types of people, and although a physical imitation of a blind man is mentioned, al-Jāḥiẓ begins and ends with a description of the imitation of a kind of speech act, beginning with funny accents, ending with a donkey imitator who imitates not the movements but the braying of a donkey. He imitates so well that he exceeds the donkeys themselves with his vocal expressions, producing what seems almost the platonic ideal of braying, a gathering of all the melodies suited to a donkey bray in a single donkey bray, the hearing of which revives the spirits of all the donkeys in audience. And that is why, al-Jāḥiẓ informs us, the ancients claimed that man is a microcosm.

Here al-Azdī ends his quotation, having satisfactorily justified his mind-bending literary project with a passage from the father of Arabic prose himself. This project consists, it seems, of himself becoming, in a written text, the imitator of an individual Baghdadi, who in turn embodied to the imitating author the intersection of all Baghdadiness, low and high, and perhaps, in fact, of the entire world in the form of a human microcosm. This explains a great deal about the peculiarity of the text's narrative structure: for example, the contrast between the long, encyclopedic lists, which help paint a literary portrait of a grand and brimming microcosm, against the written mimesis of dialogues between Abū al-Qāsim and the other party-goers, which are vividly conversational.

The text contains colloquialisms or specialised terms so ephemeral that, in their present garbled unvowelled state, they are sometimes practically meaningless to us.[9] These may indeed represent a trace of partially lost dialects and vocabularies, but one has to suspect that Abū al-Qāsim's Isfahani audience was almost as nonplussed as we are. Thus the *Ḥikāya* appears at once inaccessible and inviting. Al-Azdī has managed to capture something lastingly familiar in his crazy-old-man character, without losing the time-specific quality necessary to make his text a mimesis of a very particular time (and Pellat states in his article that in its original sense the verb *ḥakā* 'applied exclusively to the present and could not indicate an imitation of the past'),[10] so that we do not get so much an impression of Baghdad, as an impression of having lost Baghdad.

The *Ḥikāya* everywhere contrasts metaphysical time with real time, portraying the events of the day as at once a cyclically recurring affair, in which Abū al-Qāsim greets the morning with exactly the same pious phrase at the beginning and end of the text, and performs all his actions in the iterative mode describing habitual behaviours, while at the same time, the goings-on in the story are absolutely singular; Abū al-Qāsim blows in at the door of an ordinary party like an anomaly of the universe – the archetypal unexpected guest, who changes the entire order of the party. We must ask ourselves if the events represented are as singular and unusual as they seem, or are themselves in some way the typical order of things, a representation of mythical cyclical time. These events, however, also appear to be narrated in a kind of literal real time. A colleague and I timed ourselves reading a portion of the text and calculated that we would occupy about twelve hours reading slowly, with time for questions and bathroom breaks, the entire text from beginning to end, which is what the author almost seems to suggest we do.[11]

Was the *Ḥikāya* really meant to be read out loud? One thing is certain, out-loud readings of certain portions of the text still cause Arabic speakers to blush with embarrassment (I have confirmed this with experimentation as well). Would it best be read in the privacy of one's own home, preserved on just the sort of pretty, portable, mysterious manuscript as that which is now held by the British Library in London?[12] Furthermore the dauntingly exhaustive lists of carpets, dishes, and other household objects seem to spring from the author's desire to create an impression of the fullness of an entire city, and, as with Proust's *À la recherche du temps perdu*, his quest for fullness results in a daunting read. Nevertheless even these lists are rhymed, and therefore pleasing to the ear, and one cannot deny that the tradition of the live performer, the *ḥākī* whom Moreh describes at length, pervades this text.[13] Abū al-Qāsim's lengthy diversion into horse poetry towards the beginning of the book may even be a nod to the popular performer's favourite pet, the hobby horse, which were often used as props by these performers.[14] However, these same horse poems have been otherwise explained by Mary St. Germain in her dissertation as an allusion to the *riḥla* (travel) section of a *qaṣīda*.[15] Thus, just like the speech of al-Azdī's Baghdadi friend, sometimes *faṣīḥ* (formal/eloquent) and sometimes shocking, our analysis of this text must vacillate in its focus from the highest literary traditions to the lowest form of street entertainment, from *qaṣīda* to hobby horse.

The author apologises for both the colloquial elements (the *laḥn*) and the excessiveness of his *Ḥikāya*, excusing it in part by attaching a Bedouin story to the end of his tale, which story is described as *adabiyya* (belonging to polite literature), as if to make up for what he fears the reader may perceive as a deficiency in his literary education.[16] Although this Bedouin tale is missing from our unicum manuscript, and is therefore lost to us, it may have resembled 'al-Maqāmat al-Bishriyya' which often ends the *Maqāmāt* of al-Hamadhānī (depending on the manuscript tradition).[17] Unlike the other *maqāmāt* in this collection, this *maqāma* does not tell the tale of a quick-talking urban trickster (like Abū al-Qāsim), but instead describes a seemingly pre-Islamic *ṣuʿlūk* (a desert-roaming bandit) named Bishr.[18] In a tale of swashbuckling adventure, Bishr battles a lion and a snake to earn the hand of his beautiful cousin in marriage, only to marry her instead to a mysterious warrior who turns out to be his step-son. During this classic tale of desert adventure, Bishr recites poetry with vocabulary whose obscurity reflects his native mastery of the Arabic language, the desert-dwelling Arabs being widely credited with unmatched eloquence and purity of speech. If al-Azdī produced a similar tale at the end of his *maqāma*-like *Ḥikāya*, he would have proven his literacy in the classical tradition. But his protagonist's, Abū al-Qāsim's, obscure vocabulary often reflects, on the contrary, his lapses into the urban, sub-literary colloquial, and this is why the author feels the need to apologise.[19] Having offered his (now lost) Bedouin story as recompense for Abū al-Qāsim's corrupted language, al-Azdī then further excuses himself with literary quotations, and three poems attributed to that master of obscene poetry and the literarisation of urban street-speech, Ibn al-Ḥajjāj.

Although al-Azdī's turning towards poetry for justification and support is traditional in Arabic literature, the effect of his quotations is to suggest that the *Ḥikāya* is also a mimesis – rather mocking mimesis – of *adab* (polite literature) itself.[20] The first Ibn al-Ḥajjāj poem quoted suggests that his literary offering will unfold in the ordinary way: you have to forgive all the oddities brought on by his foolishness (*sukhf*). Al-Azdī's claim that his offering is in any way ordinary seems at first glance outrageous, unless you admit with him that the most ordinary thing of all is a text bespattered with the foolishness of its author. In the second poem that he quotes by Ibn al-Ḥajjāj, he seems to offer his work in the form of a strangely upright turd, or perhaps an erection, that shocks even the light-hearted litterateurs, Bishr ibn Hārūn and al-Bustī. In the third poem, he purports, along with Ibn al-Ḥajjāj, that all his speech is merely *samar*, an evening tale or conversation. The evening is the time appropriate for outrageous conversation, if ever there is an appropriate time (which is why the *1001 Nights* is not *The 1001 Days*).[21] But al-Azdī explicitly begins his tale in the morning, when Abū al-Qāsim throws off his pious robe and says 'Good morning, and no scandal!'[22] It is a kind of scandal, a nod to the topsy-turvy ethos of *mujūn*, to begin one's *samar* in the morning light.

With such an ambiguous apology before her, the reader must ask herself what kind of reproduction of eleventh-century Baghdad is presented. The *Ḥikāya* is often regarded as a possible source of historical and cultural information, and no doubt it is a treasure-trove, but one must remember, our source is a drunken and probably insane old man. And the party itself, although it seems realistic, is hardly described at all, while the ordinary-seeming Isfahani guests mostly sit on in stunned silence.

The *Ḥikāya* has been criticised for saying too much, and for the unevenness of its commitment to realism, but if we carefully examine what the author himself claims in his introduction, perhaps we can see in this unevenness the wrestling between the individual and the literary tradition, in which Abū al-Qāsim, a wild-eyed old man, once, no doubt, the master of the Arabic language, is now mastered by it. Raving all day long until he passes out on the floor, this aged wanderer alienates his would-be audience with his irrepressibly vicious tongue, even as he draws them in with an eloquence that suggests enormous personal experience.

'Here I begin my writing, having apologised for it', al-Azdī goes on after his third quotation of Ibn al-Ḥajjāj, and ends with one more, in which he claims that he is generally shy and diffident, unless he is with the right people, in which case he lets himself go.[23] With this invitation to the reader to become his co-conspirators, al-Azdī begins his *Ḥikāya*. During the course of the day, he will touch on virtually every genre of Arabic literature, though his literary citations are often wrenched from their original context and placed in outrageous juxtaposition with one another. At every turn in this world, we will find a war staged between realism and the grotesque, among other literary opposites.

Mimesis or Mannerism?

> From Icelandic, Dutch, and old English authorities, there might be quoted other lists of uncertain whales, blessed with all manner of uncouth names.

> But I omit them as altogether obsolete; and can hardly help suspecting them for mere sounds, full of Leviathanism, but signifying nothing.
>
> Melville, *Moby-Dick*, 'Cetology'

In his *Mannerism in Arabic Poetry*, Stefan Sperl defines mannerism as a style of literature for which literature, not reality, is the correlate.[24] Contrasting two descriptions of ships by the neo-classical poet al-Buḥturī (d. 897), and the more modern Mihyār al-Daylamī (d. 1037), he shows how the first (mimetic) seems to describe an individual ship at a particular time, while the second (manneristic) seems a riddle made of metaphors alluding to ships in general. Abū al-Qāsim's speech is sometimes riddle-like, and sometimes describes individual people or events at particular times. This is because he quotes from virtually all genres of Arabic literature. Consequently, anything in his speech that seems a realistic description of his experience as a citizen of Baghdad is just as likely to be one quotation in a beautiful bouquet of quotations. Can the *Ḥikāya* constitute a classically mimetic rendering of al-Azdī's contemporary Baghdad, as many scholars seem to hope? Does the *Ḥikāya* (if we are using Sperl's elaboration of his definitions of mimesis and mannerism) display 'concord between signifier and signified, [reflecting] faith in the mimetic adequacy of language?' Or rather does it show, with 'manneristic discord … despair over [language's] inadequacy as much as delight in its potential as a creator of meanings and patterns?'[25]

As described above,[26] in his introduction al-Azdī says that Abū al-Qāsim is based on a real person, but is meant to be a mimesis of an entire city, and that man himself is a microcosm. Thus he suggests that Abū al-Qāsim represents not only a real Baghdadi man he once knew, but the entire world, or the world of Baghdad (itself sometimes characterised as a microcosm of the Islamic world).[27] Thus, since the author promises us a mimesis and a microcosm, we can well understand why many have viewed the *Ḥikāya* as being dedicated, however unevenly, to realism.[28] Nevertheless, al-Azdī's introductory description of his project is focused emphatically on language – on the language of his friend (as opposed to, for example, his actions or appearance), on the language mimicked by the mimes described by al-Jāḥiẓ, and on the language of his own text – its literary sources and linguistic flaws – so that he seems to introduce not so much a mimesis of the reality of Baghdad, as of the speech of Baghdadis, about Baghdad. It is uncertain, however, that even Baghdadi speech is faithfully reproduced, as Abū al-Qāsim's diatribes are often decidedly literary in nature – more a collection of poetic quotations, strange words, and metaphors, than a Twain-like reproduction of dialectical peculiarities.[29] The *Ḥikāya* thus provides a caricature of Baghdad, and suggests that a literary microcosm of the fantastical city of Baghdad could only be, in fact, a literary microcosm of literature itself, the city of Baghdad long imagined as a centre of cultural and literary production.

In insulting a man identified by his fellow guests as a 'joker', Abū al-Qāsim applies particular venom. 'Wouldn't you like to get to know him?' the other guests ask him. 'No, by God! Why poke a bag of shit?' he replies.[30] Soon after, when Abū al-Qāsim turns his lewd eloquence towards a beautiful boy at the party, the other guests ask him 'Do you know him?' 'Yes!' he replies 'I have known him since he was a toddler … back when he could piss but couldn't talk. This is my boy! I raised him myself! I suckled him at my own bosom', and goes on to recite an obscene poem about the boy's mother.[31] Abū al-Qāsim does not address the other guests personally or try to get to know them. He uses them as cues to spew literature and universal insults. Bakhtin similarly describes the praise and blame used by Panurge and Friar John when speaking to one another: 'Formally the praise-abuse of the litanies is addressed to Friar John and Panurge, but actually they have no definite, restricted addressee. They spread in all directions, drawing all spheres of culture and reality after them …'[32] Such universally addressed insults and descriptions are not best read as representations of an individual person, time, or place.

Abū al-Qāsim so thoroughly dominates the conversation with his abusive language, that al-Azdī, the author and narrator of the *Ḥikāya*, tends to fade into the background with the Isfahani guests. He first, however, excites our suspicion by introducing his text as an 'evening conversation' and then beginning it in the morning. It is his voice, moreover (and not Abū al-Qāsim's), that names itself as the *ḥākī* of the people of Baghdad. Abū al-Qāsim himself is the gathering point of this *ḥākī*'s impression of Baghdad. He is similar to a stereotypical persona adopted by a comedian, Andy Kaufman's 'Foreign Man', for example, a hapless stand-up comic with a strong, vaguely eastern-European accent. Such a character's speech, though it dominates the narrative, cannot be considered reliable, but rather is intended to enrich the portrait of the character being created. According to Kilito's definition of the word *ḥikāya* as the imitation of the speech of a type or of a fictional character, as when a writer writes in the voice of a character,[33] the speech in a *ḥikāya* would by definition represent a markedly subjective viewpoint.

In his article '"Focusees" of Jocular Fiction', in *Story-telling in the Framework of Non-fictional Arabic Literature*, Ulrich Marzolph describes what he calls *Kristallisationsgestalten*, characters around which certain types of anecdotes crystallise, because these anecdotes exemplify the types of things that character would typically do or say.[34] For example, comic stories might be attributed to a famous jester even if they occurred a century after his death. The Cynic Diogenes, compared elsewhere in this study to Abū al-Qāsim, seems himself to be such a character, as the sayings attributed to him did not necessarily originate with him, but were simply 'apt',[35] and indeed Cynicism itself has been read as representing not so much a *telos* (a goal or final cause) as a rhetorical stance.[36] In his account of the character Abū al-Qāsim, al-Azdī offers us a comedic impression or impersonation of the people of Baghdād, similarly language-oriented or rhetorical in import.

Baghdad the Party-crasher

> By art is created that great Leviathan, called a Commonwealth or State – (in Latin, Civitas) which is but an artificial man.
>
> Hobbes' *Leviathan*, as quoted in Melville, *Moby-Dick*, 'Extracts'

> From his mighty bulk the whale affords a most congenial theme whereon to enlarge, amplify, and generally expatiate ... Applied to any other creature than the Leviathan – to an ant or a flea – such portly terms might justly be deemed unwarrantably grandiloquent.
>
> Melville, *Moby-Dick*, 'The Fossil Whale'

Like readers who seek information about Baghdad from the *Ḥikāya*, the Isfahani guests occasionally press Abū al-Qāsim for details about his hometown.[37] His answers are usually less than informative. When they ask about his house in Baghdad, he bristles, and after revealing that he lives on Sikkat al-Jawharī,[38] describes it with obscene poetry about dens of iniquity.[39] His answers are always in some way evasive. In fact, we cannot even be sure that Baghdad is his hometown; at one point in the story he recites a poem that seems to suggest that he first lived in Isfahan and only later moved to Baghdad:

> *You ask me about Isfahan (may time yet tear it down!);*
> *the young men look like old men there, the old resemble hounds.*
> *I left when just a child, and thus escaped those sterile grounds!*[40]

Like everything the character says, however, this is a literary quotation, loosely appropriate to the topic of discussion, but not necessarily autobiographical.[41] Since, in his introduction, the author al-Azdī admits to weaving multiple sources, literary and conversational, popular and personal, into one narrative, and passing it all off as his own, it is difficult to distinguish within any given section of the text between citations and inventions of al-Azdī's. We are similarly unable to determine if Abū al-Qāsim the character is reminiscing on his life or showing off his knowledge of Arabic literature when he, for example, recounts relaxing in gardens outside of Baghdad with his literary friends, or even screaming in pain as the fleas in his bed bite him in the Baghdadi night-time.[42]

Abū al-Qāsim's Isfahani audience is eventually led to question the sincerity of his opinions on Baghdad and Isfahan, so we the readers should probably question them as well. Having occupied the first ninety-two folios of the manuscript with an almost unbroken tirade against Isfahan ('the weather is dust, the soil is muck, the water is clay', he says,)[43] punctuated by heartbreaking exclamations of longing for his beloved hometown, Abū al-Qāsim drinks a glass of Isfahani water. 'By god', he declares abruptly, 'I've been unjust to the people of Isfahan in all that I've said about them.'[44] He then launches into a panegyric on the formerly maligned city's sweet water, ample flowing air, and temperate seasons, using some of the very same phrases he used to praise Baghdad earlier in the conversation. Likewise, some of the very same strategies he used earlier to insult Isfahan, he now uses to insult Baghdad. For example, he lists ugly place-names in both cities.[45]

Abruptly, he begins talking about two men, referred to only as *fulān* and *fulān* ('so-and-so' and 'so-and-so'), eating in Baghdad. Someone asks, 'What sort of a man is *fulān*?' (indicating one of the two men, who are not otherwise named or described). 'What do you mean what sort of a man is *fulān*?' he replies. 'Is the moon hidden, that you have to ask after it?' Without otherwise identifying the two anonymous men, Abū al-Qāsim goes on to exclaim at the enormous difference between them, one like Capella, the goat star (*al-ʿayyūq*), and the other like a goat (*al-ʿunūq*).[46] These comparisons continue for several pages, devolving finally into a virulent tirade against the lesser of the two men, who is 'lower than shit and filthier than manure'.[47] At last a guest interrupts the story to ask, 'Who is this whom the old man Abū al-Qāsim is describing with such infamies?'[48] Abū al-Qāsim brushes the question aside ('What's it to you?' he asks). We learn nothing about the two men in this story, except that one is heaped with exaggerated though generic praise, and the other dragged through the mud. Since they are both referred to only as *fulān*, it is unimportant which is which.

Abū al-Qāsim then goes back to describing food recently brought to the table. This leads to the further lauding of Isfahan, for indeed food seems to improve the old man's disposition. Then he turns with satirical fury against Baghdad. 'Abū al-Qāsim', says one of the guests, 'You weren't saying anything of the sort about Baghdad before now. You were just insulting the people of Isfahan'. 'Gentlemen', Abū al-Qāsim replies, 'Those camels have passed, and their cargo was carelessness. God knows that I say,

> *Neath hardwood trees in Isfahan, on unplowed ground to tarry,*
> *More dear to me and sweet it is than all of Baghdad's berries,*
> *There nights have two sides, half is of insomnia, half fleas,*
> *To whom, as they still bite, I jump and scream and utter pleas.*'[49]

In this exchange we see how Abū al-Qāsim turns his former arguments on their heads, showing that in the world of words, everything, like a Baghdadi night, has two sides: the two men who walked into the dining hall are just *fulān* and *fulān*, and arguments can be made as passionately and eloquently for Isfahan as for Baghdad. Attacking both sides of the argument with equal vigour and eloquence, Abū al-Qāsim at once demonstrates the power of language to persuade, and its promiscuous ability to make either side of any argument seem true. By placing this story at the moment Abū al-Qāsim reverses his position on Isfahan and Baghdad, the comparison of the cities falls side by side with this generic exercise in praising and blaming.

The *Ḥikāya* represents a response to certain literary treatments of the city of Baghdad and of cities in general. Debates about the relative merits of major cities, such as Kufa and Basra,[50] or Baghdad and Isfahan, were one common form of a common literary genre that featured debates between a wide range of competing entities, such as the narcissus and the rose.[51] These debates often seem as much rhetorical exercises as expressions of the author's opinion. Van Gelder, in discussing this debate style in general and the debate between Kufa and Basra in particular, suggests that 'one should, perhaps, imagine a reservoir of many statements and judgments, positive or negative, concerning the two towns, that occur as quotations in written texts, either separately or combined to form texts with direct confrontation of opposing views ...'[52] This reservoir, he explains, often exists prior to the debate itself, adding that 'it happens sometimes that a particular point is mentioned first with one town, then with the other' (for example, in the debate that he examines, both Kufa and Basra are at some point referred to as 'the dome of Islam').[53] This is to say that such portrayals of individual cities are often very much derived from literary traditions about those cities or about cities in general.

In his article, 'Baghdad in Rhetoric and Narrative', Michael Cooperson explores the dual portrayal of Baghdad in a long history of literature about the city, as at once the centre of culture as well as of debauch, both longed for and reviled, ever absent in time or space, but always present in prose or poetry. He uses Abū al-Qāsim's speeches as examples of widespread tropes, such as expressions of homesickness for Baghdad, as well as tirades against its foul smells and dangers. These last he attributes partly to the 'disillusionment' felt by the visitor to Baghdad, due to 'the city's failure to live up to such inflated expectations' as those presented in the copious literature about it.[54] Descriptions of the city, he explains, are shaped by an interaction between literary perceptions and certain 'actual circumstances of urban life', especially the tension between public and private domains within an extremely diverse community.[55] Given this intimate interaction between a literary tradition of a city and the experience of its inhabitants and visitors, readers should consider the anxiety of influence weighing on any author of literature about Baghdad.

What the *Ḥikāya* certainly reveals, however, is something about the city's reputation, a city represented as a loud-mouthed party-crasher of extraordinary verbal ability. Baghdad dominates the conversation in the *Ḥikāya*, as in medieval Arabic literature. Baghdadis themselves had a reputation for overweening pride, according to Andalusian travel-writer Ibn Jubayr (d. 1217), who writes, 'Every one of them imagines that, compared to his town, the entirety of existence shrinks into insignificance.'[56] Certainly this is an apt description of the loud-mouth party-crasher who talks all day.

As for Isfahan, the city in which the *Ḥikāya* actually takes place, the author of the *Encyclopaedia Iranica* article on the town describes it as 'second in prominence only to Baghdad'. He cites the *Ḥikāya* as his source for this claim,[57] though indeed the *Ḥikāya* seems an even poorer source of information about Isfahan than about Baghdad. In the first half of the text it is portrayed as a world of shit inhabited by monsters, and in the second half, a magical land of musk and jewels and breezes scented with wine (a mere rehashing of previous descriptions of Baghdad). Isfahan did share Baghdad's dual reputation as a place of power and luxury, and a place that had also suffered from the ravages of bad politics and natural disasters.[58] It is interesting to note that according to Yāqūt's thirteenth-century *Muʿjam al-buldān* (*Encyclopedia of Places*) the Isfahanis were proverbial for their miserliness, and the misers of Arabic literature are always the arch enemies of the party-crashers.[59] Monographs on miserliness and party-crashing tend to appear side by side,[60] so perhaps Baghdad the party-crasher was sent to Isfahan for this very reason – to face his mortal literary enemy.[61]

Yāqūt's description of Isfahan also notes the necessity of manure to render the hard Isfahani soil fertile,[62] and Abū al-Qāsim does mention this practice, saying, 'They carry their shit piled up on their heads and on the backs of their beasts of burden to their orchards, where they pollute the rivers with it and raise their crops on it, and so they eat it, yes, by my life! It is their excrement; it issued from them, and to them it returns. They are the most deserving of it!'[63] Yāqūt in fact recounts a tale that he heard from 'a merchant' (and one is tempted to call it an urban legend), in which a rich Isfahani gives people food under the condition that they evacuate their bowels on his farmland.[64] This story combines the two Isfahani traits of miserliness and the agricultural use of manure.

Yāqūt also emphasises the delicious sweet water of Isfahan's river,[65] and indeed it is after drinking a glass of Isfahani water that Abū al-Qāsim begins praising rather than insulting the city. However, the party-crasher earlier describes this river, Zandarūd, in strongly unflattering terms:

> I only see a piddling little stream in a wasteland, like a flow of poor-man's pee. When the river-level is up, it's filled with mud and garbage. When it's down, it's just a trash-heap with dung and bits of dust. They named it, in their stupidity, 'Zandarūd', meaning 'river of life'. When they're waxing poetic, they call it 'Zarīnrūd', meaning 'River of Gold'. May God make their minds go! And heat up their eyes! Were this water-course you boast of in Iraq, it would not be considered sufficient for two little hamlets. It wouldn't water two farms![66]

Abū al-Qāsim goes on to mention other Isfahani locations by name, and to provide false scatological etymologies for these names. Despite the fact that these are lists of real place names, they are used here only as the set-up for rude jokes. As for the descriptions of Isfahan as relying on manure, this may be an ancient trope; the description of Persians as 'eaters of dung', a reference to their use of manure in farming, can be found even in Herodotus.[67]

The phrases used to describe both Isfahan and Baghdad in the *Ḥikāya*[68] are often widely-used tropes; the breeze of musk or wine, for example, or the pebbles made of jewels, are metaphors also used to describe Damascus in poems by the Damascene, Ibn ʿUnayn.[69] By using these phrases to describe first one city and then its competitor, Abū al-Qāsim damages the credibility of his descriptions. In between his fanciful literary exclamations, he lists types of textiles, dishes, fruits, spices, drinks, carpets, and houses supposedly found in Baghdad. However, these are all presented as lists of things *not* found in Isfahan,[70] and the opposing lists of things found in Isfahan are so fantastically faecal in content as to make his entire encyclopedic enterprise seem unlikely to describe reality objectively (did the people of Isfahan really get in brutal street fights while vying over sewage, for example?)[71]

Just as a city must include some method by which to evacuate waste, so the obscene functions of the human body have a role in a microcosmic portrayal of an entire city, and obscene verse in an encyclopedic portrayal of Arabic literature (one is reminded of Ibn al-Hajjāj's famous verses in which he asks, 'Who could live in a house with no toilet?')[72] But the *Ḥikāya* often seems a linguistic encyclopedia – a collection of outrageous words – rather than a material one. Abū al-Qāsim may be listing 'lexical rarities' (*ghurabāʾ*)[73] and swearing encyclopaedically. As for the poems he recites, these tend to show us little about Baghdad or Baghdadis, except, perhaps, for their predilection for quoting copious verse.

To characterise the city of Baghdad as a party-crasher, who habitually drops in on gentlemen's parties, and then repeatedly hints that he is hungry, also emphasises the greedy consumerism of the capital. Indeed it is a topos of banquet literature to characterise the empire as a body that consumes.[74] In ancient Roman literature, a dish called the 'Shield of Minerva' included ingredients from all over the Roman domain, so that the consumer could imagine eating the whole world in one platter.[75] Abū al-Qāsim's enthusiastic lists of foods likewise include many dishes named for far-flung places around the Islamic world, so his depiction of dining in Baghdad creates an impression of eating the entire Islamic empire ('Kaskarī duck and Ṣarṣarī kid, and Indian fattened chicken, and Turkoman suckling lamb!')[76] Indeed one of Baghdad's detractors described it as a terrifying consumer, writing that, 'during the time of her dominion, [it was as if] an insatiable bloodsucker had swallowed up the whole world.'[77]

Abū al-Qāsim's greedy world-devouring rants bring to mind the narcissistic host in the Petronius' *Cena Trimalchiones* (Trimalchio's Dinner-Party), who serves a series of absurd dishes designed to overawe his guest. One course depicts the entire zodiac, thus inviting the diner to eat the universe itself. Catherine Connors argues that, with this grandiose dish, Trimalchio portrays the scope of his own doings as 'imperial, drawing on the idea that Roman *imperium* stretched over sea and land.'[78] Elsewhere the *Satyrica* makes this point more explicitly through Eumolpus's poem on the Roman Civil War, which begins, 'The victorious Roman now held the whole world … nor was he satiated.'[79] It goes on to describe all the corners of the world from which Rome ravenously gobbled exotic goods and beasts, for entertainment and for food. Trimalchio himself, though he makes himself a very symbol of this imperialist greed, also recites a poem chiding Rome for its world-devouring nature.[80] This poem is traditional in its moralising theme, but in the mouth of greedy, extravagant Trimalchio, its moral content is difficult to digest without a grain of salt.[81] Likewise the *Ḥikāya* contains multiple traditional literary stances for and against Baghdad, but they are all fatally undercut by their context. Abū al-Qāsim's moral advice in particular, when he is asked how to be a gentleman, for example, takes a comic dive straight into the gutter.[82]

The empire, when thus characterised as a greedy host of dubious morals, rarely offers digestible sustenance to its guest – some Roman examples involve a garish host who, either by his stinginess or by his nauseating display, fails to satisfy anyone.[83] Trimalchio himself is an intruder on high-society and power, a party-crasher of his own feast. But Abū al-Qāsim al-Baghdādī is not the host of his feast at all, and offers nothing more than words to consume, stomach-turning or otherwise. Indeed he is like the party-crasher described in al-Khaṭīb al-Baghdādī's eleventh century *al-Tatfīl* (Party-Crashing):

> *More pushy than nightfall, and faster . . .*
> *If he's in the house, he's the master!*[84]

He does not own the house, but he dominates it by means of his irresistible, attention-arresting verbosity. At one point he even refers to the event he crashes as his party, saying of a fellow guest, 'This young man, God bless him, came to my party today, but do you know what he ate?' 'No', they would say. 'A thousand cocks in a loaf of bread!'[85]

Party-crashers have no place or power at the banquet beyond their mighty silver tongues, with which they dominate all else. Abū al-Qāsim, representing Baghdad, may be a beggar at the table of Isfahan, but empty words about Baghdad are the standard against which Isfahan is measured in Abū al-Qāsim's speech. One of his chief criticisms of the city concerns its inhabitants' illiterate and sloppy conversation, which he mimics in semi-intelligible Persian. For example, Abū al-Qāsim, mocking the coarse speech of the Isfahanis, imitates a street-caller's cry, saying something in Persian that he then himself translates into Arabic with the equivalent of, 'Hey lady, I'll haul out your shit!'[86] According to contemporary anti-Arab, Perso-centric discourse, the Arabs were lizard-eating upstarts from the desert, crashing the party of former Iranian glory.[87] The Arabic language, the language of the Qur'an, and Abū al-Qāsim's double-edged weapon, is the locus of their upstart empire's new power. It is at once the language of the state, and the language in which God chose to present mankind with His final revelation. But like the famous Abbasid caliphs, often featured in literature, but remaining in Baghdad chiefly as a figurehead while the Persian Buyids there held actual political sway, Abū al-Qāsim may have literary dominance over the party that he crashes, but the meat and potatoes of the Isfahani feast are provided by the Persians, so to speak.[88]

Like Abū al-Qāsim, both beggar and master, Baghdad offered a double perspective to the observer. Many note, for example, that Baghdad could be a paradise for the rich and hell for the poor.[89] Cooperson compares the city to 'the Afterlife, where the blessed and the damned can observe each other by looking over a wall.'[90] In his *Plot of Satire*, Alvin B. Kernan describes Orwell's *Down and Out in Paris and London*, which observes first the elegance of an upscale restaurant's dining room, and then the chaos and filth of the kitchen where the food is prepared. 'The point at the soundproof door', writes Kernan, 'is the position the satiric description always occupies.'[91] Abū al-Qāsim's wide veering between paradisiacal luxuries and hellish filth gives a similar impression of a double perspective on an environment in absurd contradiction with itself.

A legendary city of power, Baghdad at once wielded its powerful hold on the cultural imagination and fell victim to it. Likewise Abū al-Qāsim dominates the party with his speech, but is ultimately dominated by it, and passes out on the floor in drunken exhaustion, having scared off and wounded any potential new friends with his sharp tongue. Like Trimalchio's party, which ends in utter chaos, Abū al-Qāsim's bombastic drunken collapse suggests the decadent reeling of power on its last legs. Of course neither Trimalchio's Dinner-Party nor the *Ḥikāya* take place in the centres of power, in Rome or in Baghdad, as if these legendary cities were always already gone and passed away.[92] Nevertheless, like Trimalchio, Abū al-Qāsim holds his audience's attention riveted to the bitter end, and presumably wakes up the next morning ready to crash another party. Both Abū al-Qāsim and the city of Baghdad may be old, but they are both perennially powerful and vigorous in the field of literary domination.

Those Camels have Passed

> … that all this should be, and yet, that down to this blessed minute (fifteen and a quarter minutes past one o'clock P.M. of this sixteenth day of December, A.D. 1851), it should still remain a problem, whether these spoutings are, after all, really water, or nothing but vapor – this is surely a noteworthy thing.
>
> Melville, *Moby-Dick*, 'The Fountain'

Theories of mimesis in literature often address the passage of time portrayed in literature. Due to the passage of time, any repetition necessarily contains difference, since it occurs later than the original.[93] Thus no mimesis can be exactly like that which it imitates, for time itself has changed it. When questioned on his fickle changes of conversational temperament, and especially his sudden reversal in comparing Baghdad to Isfahan, the old man Abū al-Qāsim replies, 'Those camels have passed, and their cargo was carelessness'. In addition to marking his literary descriptions as 'careless', this remark evokes the passage of time during the day-long party.

The *Ḥikāya* throughout evinces an awareness of the passage of time, from al-Azdī's introductory remarks on the time-frame of the story, to markings of various stages in the party, the meals, and the singing. Al-Azdī creates an unprecedentedly realistic portrayal of real time within literary time, without summarising or skipping over events. This day unfolds in a singularly realistic portrayal of time passing, but, at the same time, the narrative's use of the iterative mode describes actions habitual to the protagonist. The day itself is recurring, archetypal – therefore, on a mundane level, necessarily unreal, in the sense that a literary trope is unreal. Kennedy writes of the *Ḥikāya* in relation to the *maqāmāt* in particular that it is 'the most significant surviving example of an "iterative" story, which is to say a story that is part of a perpetual cycle of similar or near-identical episodes.'[94] The iterative mode of the narrative is partly responsible for its 'unreal' feeling, as the actions described are given only as examples of the type of thing that used to usually happen.

Bakhtin's studies of the relation between time and space in literature may help us think about the portrayal of time passing in the *Ḥikāya.* He defines the word 'chronotope' as 'the intrinsic connectedness of temporal and spatial relationships that are artistically expressed in literature.'[95] He provides many examples of common chronotopes found in literature, such as 'adventure-time', the chronotope of the typical ancient Greek romantic novel, and 'adventure-everyday' time, the chronotope of the *Satyrica* and Apuleius' *Metamorphoses,* a model relevant to this study for the clash it stages between literary conventions and 'real life'.[96] But none of the chronotopes that he describes are like the *Ḥikāya,* a real-time portrayal of a single day passing in a single setting. From the moment Abū al-Qāsim enters the party, the narrative measures the time elapsing with the unabridged content of his conversation, until the moment that he goes to sleep. That is to say, he talks all day, and his conversation is recorded in full. Thus it takes all day to read this full transcript of his conversation. While other texts depicting full conversations may use similar one-to-one ratios of plot-time to reader-time, none make the chronotope so explicit, and most freely use summaries or leaps in time often found in narratives.

However, some similar examples should be mentioned. A very recent but striking comparison can be found in the television show *24*, which, in twenty-four hour-long episodes, portrays the passing of a single action-packed day. However, it takes an entire season of TV shows to watch the portrayal of this single day. Similarly, although the *Ḥikāya* could technically be read aloud in a twelve-hour period, it took me three years to come to a provisional understanding of its dense content. This discrepancy in both cases only serves to emphasise how very action-packed and wearisome the day represented was.

Texts that divide stories into days and nights, like the *1001 Nights* and the *Decameron*, provide another interesting point of comparison. Al-Tawḥīdī produced a similarly organised non-fictional text, *al-Imtāʿ wa-l-muʾānasa* (Book of Enjoyment and Good Company),[97] which portrays thirty-seven evenings of conversation, the topics divided by night. The *Ḥikāya* does not, like many contemporaneous Arabic texts, contain individual *akhbār* (anecdotes), each proceeded by an *isnād* (chain of transmission), but rather seems to attribute the entire text to al-Azdī at the beginning of the introduction, as though what follows represents a single, extremely lengthy *khabar* (anecdote) – a single day or night, in extreme close-up, taken from a chain of stories in which Abū al-Qāsim habitually crashes parties. A fractal-like effect is achieved, in which a single day can be shown to contain the complexity of an entire universe, just as a single *khabar* hypothetically may contain enough material to fill an entire book. Here we find one more effect – that of the zoom-in close-up – with which to achieve the feeling of the microcosmic fullness of this single book, or banquet, or man. Nicholson Baker's *Mezzanine* may provide a further fruitful point of comparison, portraying the thoughts of a man riding an escalator from the ground floor to the mezzanine level in such detail as to occupy the entire length of the novel, showing that an event that may occupy only a small portion of a typical narrative could, if closely examined, itself provide material for an entire book.

It is difficult to guess how al-Azdī intended his book to be read, or over what period of time. To try to read it at one sitting, as is seemingly suggested in the introduction, is to wonder, with Kennedy, if the text were not intended to be deliberately wearisome.[98] By thus exhausting the reader, the narrative draws attention to the passing of time in a narrative relative to the experience of time passing as a reader of a narrative.[99] What is it like to read a book that unfolds at the pace of the narrative itself? The *Satyrica* matches the passage of time during a journey with the length of a poem recited by one of the journeyers.[100] The tedious length of the epic poem recited by Eumolpus matches the tedious length of the journey to the city of Croton. The poet's fellow travellers carry heavy baggage and listen to Eumolpus recite throughout the painful hike. Similarly, the tedious length of Abū al-Qāsim's speech emphasises the sensation of a day passing, minute by (occasionally excruciating) minute.

Mez considers literary treatments of boredom (*le genre ennuyeux*) as crucial to the development of the literary atmosphere in which the *Ḥikāya* was written. He quotes Jaʿfar, the Barmakid vizier, commanding a poet to 'Sing and be brief!' (*Sing und mach's kurz!* | *qul abyātan wa-lā tuṭil!*). As for prose, he writes, 'nothing was abhorred so much as a long story.'[101] The introduction to the *Ḥikāya* addresses this concern:

> For those readers who are eager and have the energy to hear this work out, and who do not find its long-windedness or literary overflowings to be a strain on their minds, and who do not consider the common expressions evidence of my faulty vocabulary or think less of me for using them, (especially when they arrive at the literary Bedouin imitation that I added to the end, and if they follow the saying that 'the spice of wit is low language, its charm is brevity, and its vitality in keeping it short'), for such readers I have gone to great pains, and have burdened myself so that they are left with what is enjoyable.[102]

The translation of this last phrase (*kallaftu lahu min al-basṭ jahdahu al-mutʿib ʿalayya wa-ghayrhu al-mumtiʿ lahu*) is debatable. Gabrieli declares the sentence obscure to him, but offers this interpretation: 'And to this kind listener

I shall be greatly indebted for having bestowed on me his undivided attention to the point of exhaustion.'[103] His translation puts all the burden of exhaustion on the reader. St. Germain gives: 'To him I have dedicated myself to (creating) an elaboration exhausting to me when some other option would have pleased him more.'[104] Her translation suggests that the author, even knowing that some other way would have been more pleasing, took great pains to produce something deliberately wearisome (as Kennedy suggests). My translation is based on the opinion that the sentiment expressed might be similar to that of al-Jāḥiẓ on the virtues of the book:

> … Anyone who chances upon a compendious (*jāmiʿ*) book … will reap the profits (*ghunm*), whereas its composer (*muʾallif*) will have to bear the costs (*ghurm*): his [i.e. the reader's] will be the benefit, while the toil and distress will be its author's (*ṣāḥib*).[105]

Al-Jāḥiẓ makes this statement in the introduction to his *Book of Living*, a work that is in its own way, like the *Ḥikāya*, encyclopedically, bewilderingly self-contradictory. It would be untrue in either case to claim that the reader is not burdened by reading these books. In fact, we should not rule out the possibility that this sentence in the introduction of the *Ḥikāya* is intended to be obscure or slippery. It might be the first sentence of the text that we cannot understand, but it certainly is not the last; the inscrutability of its language is one of the distressing or tiresome features of this work, and the introduction apologises for it in distressingly obscure language.

In concluding his description of the work, Mubārak notes the extreme thoroughness of its literary *mujūn*: 'He left no door of jest un-knocked upon … I think he crammed his book with the most filthy dissolute poetry ever recorded.' In noting its exhaustive treatment of the subject, Mubārak effectively acknowledges the occasional tediousness of this text. He then adds, 'This type of composition has value anyway, for it is the sort of literature that the soul requires in times of boredom.'[106] Thus he sees it nevertheless as not boring, but indeed as a relief from boredom because of the spiciness of its subject-matter. This echoes a sentiment commonly expressed in medieval Arabic literature, as, for example, in hadith scholar al-Khaṭīb al-Baghdādī's *Art of Party-Crashing* (*Kitāb al-taṭfīl*, discussed below in Chapter 3):[107] "ʿAlī,[108] may God be pleased with him, said,

"If your minds get tired, just as bodies do, seek out some entertaining information!"'[109]

Although readers have found the *Ḥikāya* and its protagonist both shocking and entertaining, it is difficult to deny that Abū al-Qāsim talks too much and can consequently grow tedious. The Isfahani audience decides that 'that's enough of Abū al-Qāsim and his conversation', very shortly after his arrival at the party.[110] Moreh's reading of the work as a theatre script may be an attempt to rescue it from its own long-windedness, as he writes that,

> Apparently Abū al-Qāsim did not perform all the *ḥikāya* at one time, but rather used bits suitable to various occasions. The author-redactor unified them in a single *ḥikāya* to be recited or acted by an actor who was free to improvise as he saw fit in response to the reaction of the audience.[111]

Moreh's interpretation raises the question of how we are meant to read the *Ḥikāya*. If we pick and choose from the text as he suggests, we can avoid the tediousness of Abū al-Qāsim's tyrannical style of conversation. But if we are to experience it in full as the author suggests in the introduction, we, like the Isfahani audience, are doomed to suffer some frustrations; we will feel what it was like to be at the party. In *The Loaded Table*, Emily Gowers, in discussing Martial's epigrams, suggests that 'the limits of Martial's frankness are nebulous, as though the way we read his poems depends on how drunk we are',[112] arguing that various poems or books in his collection themselves represent dinner parties, or the stages of drunkenness achieved during such a party.[113] If the *Ḥikāya* was intended to be read out loud at one sitting to an audience, we should certainly hope that this audience would be intoxicated, for their sake.

In his *On the Art of Medieval Arabic Literature*, Andras Hamori describes the wine poem's intoxicated focus on the present moment, and how thereby 'the libertine subdues time', at least for a moment.[114] The famous poet Abū Nuwās (d. 814), for example, plays with the traditional theme of the *aṭlāl*, the traces left in the sand by the departed beloved's desert encampment. Pre-Islamic and early Arabic *qaṣīda* poems tend to begin with the narrative voice mourning over these traces and the bygone romantic joys they evoke.[115] Abū Nuwās's poetry doesn't exactly ignore this traditional nostalgic pain, but rather drowns that pain in alcohol.[116] Late in the party, when Abū al-Qāsim begins to grow seriously drunk, he suggests that the party guests continue drinking in the morning. Like his disavowal of his old age, his drunken reluctance to retire as the party draws to a close shows Abū al-Qāsim as one who strives to transcend the natural course of time. One is reminded of Yaḥyā al-Barmakī's poem to his son al-Faḍl,[117] in which he articulates several popular topoi of Arabic literary discourse, here translated by Reynold Nicholson:

> *Seek Glory while 'tis day, no effort spare*
> *And patiently the loved one's absence bear;*
> *But when the shades of night advancing slow*
> *O'er every vice a veil of darkness throw,*
> *Beguile the hours with all their heart's delight:*
> *The day of prudent men begins at night.*[118]

Like the passage of the seasons, or stages of youth and old age, different times of day are appropriate for different types of behaviour. Similarly, at specific times of the year devoted to carnivalesque festivities, topsy-turvy behaviour is the norm.[119] But at the party crashed by Abū al-Qāsim, even the normal boundaries enforcing the practice of foolery are reversed. As al-Nīsābūrī (d. 1015) clearly explains in his eleventh-century *ʿUqalāʾ al-majānīn* (Wise Fools),[120] there are certain appropriate times for foolishness (e.g. the evening, or during one's youth). Venerable white-bearded old age, which is Abū al-Qāsim's stage in life, is explicitly not one of them.[121] Furthermore, instead of occurring on a carnival day or even just during the night-time, the *Ḥikāya*, describing actions throughout as habitual (e.g. 'Abū al-Qāsim would walk into a party …'), presents a recurring and potentially ceaseless descent and redescent into the inappropriate and strange.

Al-Azdī writes in his introduction, 'This *ḥikāya* amounts to the events of one day, from its beginning to its end, or likewise, a night.' Al-Shāljī understandably emends this to 'the events of one day … *and* likewise, a night', (emphasis mine) because the original is more or less nonsense: we know that the story begins in the morning, and lasts into the night. There is no question of it being either a day *or* a night, and the usage of 'likewise' (*kadhālika*) with 'or' (*aw*) constitutes a kind of subtle bit of nonsense in both Arabic and English. However, this nonsensical little word may represent not the error of a (usually careful) scribe, but the subtle introduction of a portrayal of topsy-turvy time. Perhaps what the *Ḥikāya* represents is a walrus-and-carpenter day, in which:

> *The sun was shining on the sea,*
> *Shining with all his might . . .*
> *And this was odd, because it was*
> *The middle of the night.*[122]

In fact, late in the day, Abū al-Qāsim recites a similar poem:

> *We went out early, at dawn, in the night, in the evening, after midday,*
> *And hunted rabbits, jackals, wolves, but donkeys got away!*[123]

Abū al-Qāsim recites this poem, discussed further in Chapter 4, while he is playing chess. As described there, this chess game brings the unusual effects of the narrative's tense and style into sharper focus, as each chess move is described only as a hypothetical move that could have been made in a habitually recurring game, rendering the specific game played impossible to reconstruct.

Jorge Luis Borges's short story 'The Book of Sand' tells the story of a book lover who acquires a book with infinite pages that, like the sand, has no beginning or end. As the Bible seller who sells the book to him remarks, 'If space is infinite, we are anywhere, at any point in space. If time is infinite, we are at any point in time.'[124] By telling his tale in the iterative mode, by ending his tale the same way it began, and by introducing his tale as a microcosm, the author creates a similarly baffling portrayal of time in a text. The day is at once today, every day, and no day. Compare this with the works of Rabelais (which are likened to the *Ḥikāya* below in Chapter 5),[125] in which 'the feast at the beginning of *Gargantua* has a precise date: it takes place on 3 February in the middle of the Carnival . . .'[126]

In conclusion, the *Ḥikāya* represents a party in which the guests appear to grow hungry and desirous of drink at realistic intervals, as if a day is actually passing. The reader grows weary and exhausted with the party guests during Abū al-Qāsim's day-long tirade, as if she herself were another bewildered member of the feast, overwhelmed by the party-crasher. Despite these gestures towards hyper-realism, this time is also lent a cosmic quality by the cyclically repeating form of the story, and what seems at first a normal, realistically portrayed day, may be a very topsy-turvy sort of day indeed. Thus the structure of the text itself seems at first to mimic reality – the tedious passing of an actual twelve-hour sitting. But like other seeming gestures towards realism in the *Ḥikāya,* closer examination reveals it as a parody of passing time and its portrayal in literature – a day neither real nor typical, but rather an enactment of literary monstrosity, and the pleasure and pain that it causes.

The author's introduction to this portrayal of a day-long party raises several important points. It tells us that it is a collection of citations, mixed indiscriminately with the author's own writing, offered as an imitation of the speech of a friend who in turn represented Baghdad to the author. The individual and the microcosm are shown to be mixed in a complex gesture of literary mimicry. The introduction also tells us that although this protagonist is based on an individual man, he represents all the citizens of Baghdad, high and low, in their total essence. We are led to ask why the author would choose a party-crashing character like Abū al-Qāsim to represent this city and what kind of information about the city of Baghdad the *Ḥikāya* might actually contain. Finally, the introduction tells us that the text can be read in the same amount of time that the events portrayed took to occur, our first hint of the importance of time passing as portrayed in this work. Our exploration of this theme once again leads us to question the *Ḥikāya*'s representation of literature's ability to represent daily life.[127]

Notes

1. See pp. 31–4.
2. *Ḥ*, p. 42.
3. Pellat, 'Ḥikāya', p. 367.
4. Abū al-Qāsim shares his *kunya* with the prophet Muḥammad, and like so many trickster figures in world literature, he appears to have certain spiritual qualities. Indeed the *Ḥikāya* itself in certain respects resembles the Qur'an, as shown in Chapter 5, pp. 155, 158.
5. See the section 'Scholarship' in the introduction (pp. 8–9), which explains the importance of the translation of the word *ḥikāya* to the history of this text's reception.
6. Abdelfattah Kilito probably best defines the word in this context as the imitation of the speech of a type or of a fictional character (*Séances*, p. 157), as when

a writer writes in the voice of a character. For more on this, see 'Mimesis or Mannerism' in Chapter 2 below, p. 77.

7. Also 'form' or 'image.' It was translated as 'paradigm' in Azarnoosh, 'al-Azdī'.
8. He elaborates in his *Book of Living*, or *Kitāb al-ḥayawān. Al-Jāḥiz: In Praise of Books* looks to Q 29:64 for this sense of the word *ḥayawān* (p. 9). It has previously often been translated as the *Book of Animals*.
9. For example, the section on Baghdadi sailor's slang, which can be found in *Ḥ*, pp. 318–19, is discussed in Chapter 3, p. 114.
10. Pellat, 'Ḥikāya', p. 368.
11. My colleague was Hassan Hussein, now lecturer in Persian at Columbia University.
12. At the eighteenth annual meeting of the Association of Literary Scholars, Critics, and Writers, Amy Richlin, participating with me on a panel entitled 'The Rules of the Writing Game: History and Fiction from Vergil through Geoffrey of Monmouth and the Arthurian Vulgate Cycle, to Gibbon and Carlyle', made the intriguing suggestion that the novel and similar written narratives originated almost as plays or public performances that could be enjoyed alone in the privacy of one's own home.
13. Moreh, *Live Theatre*, pp. 94–103. See the section 'Scholarship' in the introduction (pp. 16–17).
14. These toy horses and the farces performed with them are described in Moreh, *Live Theatre*, 'Players of *Kurraj*', pp. 27–37.
15. St. Germain, *Anomalous*, pp. 53–77. A *qaṣīda* is a particular form of poem associated with pre-Islamic composition, described in the 'Translations' section of the introduction (pp. 6–7).
16. *Ḥ*, p. 44.
17. Al-Hamadhānī, *Maqāmāt*, pp. 250–8.
18. A *ṣuʿlūk* was a solitary desert rogue who lived by his own rules, the author of a particular genre of brigand-themed poetry.
19. Abū al-Qāsim's speech actually includes instances of both types of obscurity, as indeed he quotes from pre-Islamic poetry not unlike that recited by Bishr. Nevertheless, Kilito names the *Ḥikāya* alongside the poetry of Ibn al-Ḥajjāj as being 'truffled' with vulgar linguistic oddities (*Séances*, p. 63).
20. Kennedy's useful definition of *adab* is cited in the 'Scholarship' section of the introduction (p. 16), where his essay 'The *Maqāmāt* as a Nexus of Interests' is discussed.
21. *Asmār* (the plural of *samar*) are coupled with *khurāfāt* (fairy tales and talking

animal fables) in Ibn al-Nadīm's famous tenth-century *Fihrist,* or catalogue of books. This category includes an early description of the *1001 Nights* (Ibn al-Nadīm, *Fihrist,* 470).

22. *Ḥ*, p. 56.
23. *Ḥ*, pp. 45.
24. Sperl, *Mannerism,* p. 164.
25. Sperl, *Mannerism,* p. 180.
26. See p. 71.
27. See Kraemer, *Humanism* VII, and Cooperson, 'Baghdad in Rhetoric', pp. 99–113.
28. See also Gabrieli, 'Sulla *Ḥikāyat*', p. 35.
29. Kennedy calls it 'a repository of *gharāba*' ('*Maqāmāt* as a Nexus of Interests', p. 165). Al-Shāljī does note that some of the words and phrases used in the *Ḥikāya* are still used in Baghdad today, especially during a section, quoted in Chapter 1 (pp. 44–6), in which Abū al-Qāsim quotes a letter by the slave girl Zād Mihr (*Ḥ*, pp. 230–8).
30. *Ḥ*, p. 75.
31. *Ḥ*, p. 79.
32. Bakhtin, *Rabelais,* p. 419.
33. Kilito, *Séances,* p. 157. Montgomery similarly writes that the *ḥikāya* as portrayed by al-Jāḥiẓ 'combines the idea of mimesis and mimicry with that of verbatim and accurate quotation, almost to the point of impersonation or ventriloquism', *al-Jāḥiẓ: In Praise of Books,* p. 153.
34. See also Kilito, *Author,* p. 63.
35. Krueger, 'The Bawdy and Society', p. 224.
36. See Branham's 'Defacing the Currency'.
37. This line of questioning begins with them asking the meaning of the word *dādhī,* which he uses in insulting Baghdad, saying the wine is undrinkable without *dādhī.* To this question he supplies a vague but semi-helpful answer, implying that it is a good-smelling thing put in date wine (*Ḥ*, p. 312). Dozy identifies it as St. John's Wort, and describes how its seeds were used in Baghdad to increase the odoriferousness of date wine (*Supplément,* p. 419).
38. 'Alley of the Jeweller'. I have been unable to identify this street in, for example, Le Strange, *Baghdad.* Maybe his answer is a joke of some kind.
39. *Ḥ*, p. 322. Translated in Chapter 1, pp. 58–9.
40. *Ḥ*, p. 90.

41. This poem is apparently by Abū al-Faḍl al-Muẓaffar ibn Aḥmad al-Yazdī, a physician and poet, who returned to Isfahan in the late eleventh century (placing the probable composition of this poem at a later time than the generally accepted composition of the *Ḥikāya* in the early eleventh century). It is, however, possible that al-Yazdī did not in fact compose the lines quoted in the *Ḥikāya*, which could have derived them from an earlier source. See al-Qifṭī, *Ikhbār al-ʿulamāʾ*, (also known as *Tarīkh al-ḥukamāʾ*), p. 215 and al-Kātib al-Isfahānī, *Kharīdāt al-qasr*, pp. 154–60. Many thanks to van Gelder for this reference.
42. *Ḥ*, pp. 268, 308.
43. *Ḥ*, p. 91.
44. *Ḥ*, p. 300.
45. *Ḥ*, pp. 93–4, 310–11.
46. *Ḥ*, pp. 301–2. A similar passage is translated in Chapter 1, p. 59.
47. *Ḥ*, p. 304.
48. *Ḥ*, p. 305.
49. *Ḥ*, p. 308.
50. See van Gelder, 'Kufa vs. Basra', pp. 339–62. Also see Mez's introduction to his edition, *Abulḳâsim*, p. xvii.
51. See Heinrichs, 'Rose vs. Narcissus', pp. 179–89. Similar debates are found even in Herodotus, as described by James Romm's study which, he writes, 'focuses on cultural encounters in which an alien perspective is privileged over that of a native' (Romm, 'Dog Heads', p. 123). Narratives, for example, in which a barbarian mocks civilised life (p. 135) themselves resemble another medieval Arabic literary topos in which a Bedouin Arab encounters civilisation, often exposing its culture as comparatively miserly. Sometimes the Bedouin seems stupid in comparison to the city-dwellers, and sometimes more noble. Romm argues that these narratives of encounters with strangers from other civilisations can often reveal the illogical nature of one's own civilisation, the same experience as encountering a Cynic, whose contempt for social decorum and customs can often remind us of Abū al-Qāsim and his trickster ilk. Abū al-Qāsim does not, however, reveal the flaws of the Isfahani society in which he is a stranger (though he insults them for a large portion of the book), but rather, with his rudeness, reveals his own flaws and, by analogy, the perceived flaws of the Baghdadi people.
52. Van Gelder, 'Kufa vs. Basra', p. 343.
53. Van Gelder, 'Kufa vs. Basra, p. 344.

54. Cooperson, 'Baghdad in Rhetoric', p. 102, also mentioned in the section 'Mimesis or Mannerism?' above.
55. Cooperson, 'Baghdad in Rhetoric', p. 111.
56. Cooperson, 'Baghdad in Rhetoric', p. 101. Von Grunebaum characterises this prideful boasting as a topos of urban literature, quoting Broadhurst's translation of Ibn Jubayr's descriptions of the inhabitants of Baghdad: 'As to [Baghdad's] people, you scarce can find among them any who do not affect humility, but are yet vain and proud. Strangers they despise ... Each conceives, in belief and thought, that the whole world is but trivial in comparison with his land, and over the face of the world they find no noble place of living save their own ... They trail their skirts trippingly and with insolence, turning not, in deference to God, from that of which He disapproves ...' ('Aspects', p. 264).
57. Kamaly, 'Isfahan'. The page number he provides for this citation (p. 17) in fact refers us to Mez's introduction to the *Ḥikāya*, which in turn cites Ibn al-Faqīh (a ninth–tenth century Persian geographer), who, according to Mez, called Isfahan the 'second Baghdad'.
58. Kamaly, 'Isfahan', passim.
59. Yāqūt, *Buldān* IV: 985.
60. See, for example, Rosenthal, *Muslim Historiography*, p. 432.
61. Mez suggests that behind this literary opposition of cities there was the historical circumstance of rebellion in the east of the Abbasid empire, as well as a rebellion in 932 by Isfahani residents of Baghdad, in which they disrupted a sermon, attacked the preacher, and pelted the mosque with stones (al-Azdī, *Sittenbild*, p. xvii).
62. Yāqūt, *Buldān* I: 294.
63. *Ḥ*, p. 91
64. Yāqūt, *Buldān* I: 294.
65. Yāqūt, *Buldān* I: 294.
66. *Ḥ*, p. 109.
67. Romm, 'Dog Heads', p. 125.
68. *Ḥ*, p. 301.
69. Von Grunebaum provides a useful collection of such poems in his 'Aspects', pp. 262–3.
70. Van Gelder made this observation after my presentation of Man Behind the Mouth.
71. *Ḥ*, p. 110.
72. Antoon translates, 'Can there be a house with no privy | and can a sane man

live in it?' (*Poetics of the Obscene*, p. 13). Bryan Turner notes that the close quarters necessitated by an urban environment inevitably leads to the obtrusion of bodily functions and the obscene on the psyche of its inhabitants, and indeed the *Ḥikāya* falls within a tradition of satirical literature that couples an urban environment with a discourse of filth and obscenity. Cf. Juvenal's famous third satire on the city of Rome, and John Gay's poem 'The Walker through London' (*Body and Society*, p. 118 ff.).

73. Kennedy, '*Maqāmāt* as a Nexus of Interests', p. 165.
74. Gowers, *Loaded Table*, p. 12. Although the following comparisons are literary rather than historical, Patricia Crone reminds us in her 'Imperial Trauma: The Case of the Arabs', of the historical similarities between the Roman Empire and the early Arabic empire 'as a result of their expansion [...] namely how to prevent the ways of the conquered peoples from undermining those of Rome' (pp. 107–16).
75. 'The Shield of Minerva, which contained all the products of the world in miniature, summed up and trivialized supreme power' (Gowers, *Loaded Table*, p. 207).
76. *Ḥ*, p. 157.
77. Cooperson, 'Baghdad in Rhetoric', p. 102, citing the thirteenth-century Armenian chronicler Kirakos of Gandzak as described in Boyle, 'The Death', pp. 145–6.
78. Connors, *Petronius the Poet*, p. 113.
79. Petronius, *Satyricon* 119: 1.
80. Petronius, *Satyricon* 55: 6.
81. Connors, *Petronius the Poet*, p. 56.
82. 'True studs, they say, do not demur | so climb on filth, and fuck a cur!' he replies (*Ḥ*, p. 83).
83. Cf. Juvenal's *Satire* 5 and Horace's *Satire* 2.8. Al-Hamadhānī's 'al-Maqāmat al-maḍīriyya' (discussed below, p. 106) provides an intriguing Arabic comparison.
84. Al-Khaṭīb al-Baghdādī, *al-Taṭfīl*, p. 75. All translations of this volume are from Selove, *Art of Party-Crashing*.
85. *Ḥ*, p. 290. See the translation of the chess game in Chapter 1, p. 55.
86. *Ḥikāya*, p. 22. Al-Shāljī renders this cry (here unvowelled) as *Ay zn bwākht kshm* (*Ḥ*, p. 110). In her translation, Mary St. Germain suggests this reading in a footnote: *Ay dibānuā guhat kisham* (*Anomalous Text*, p. 203).
87. Van Gelder, *Dishes*, pp. 25, 30–3, and Crone, 'Post-Colonialism', p. 17.

88. Potatoes, domesticated in the Americas, were not yet part of Eurasian cuisine. See the 'Scholarship' section in the introduction for more on the Buyids (p. 8).
89. The narrator of *Daʿwat al-aṭibbāʾ*, who is himself a kind of party-crasher, claims he left Baghdad for this very reason (Ibn Buṭlān, *Daʿwat al-aṭibbāʾ*, p. 53). Ibn Buṭlān, d. 1038, was a Christian physician and writer who is discussed further in Chapter 3, p. 104.
90. Cooperson, 'Baghdad in Rhetoric', p. 106. This observation is in reference to the letters of slave Zād Mihr to her master, in which she complains that 'Baghdad is a paradise for the rich and a torment for the poor' (translation Cooperson's).
91. Kernan, *Plot of Satire*, p. 86.
92. Trimalchio's dinner party is held on his estate in Campania, in southern Italy.
93. The question of mimesis-as-repetition as well as an overview of such theories concerning mimesis and the passage of time (by such scholars as Erich Auerbach, Walter Benjamin, and Jacques Derrida) can be found in Melberg, *Theories of Mimesis.*
94. Kennedy, '*Maqāmāt* as a Nexus of Interests', p. 162. Gérard Genette describes this tense in his *Narrative Discourse* as 'where a single narrative utterance takes upon itself several occurrences together of the same event (in other words … several events considered only in terms of their analogy)' (Genette, *Narrative Discourse*, p. 116). He focuses especially on Proust's innovative use of the iterative mode in his *À la recherche du temps perdu.*
95. Bakhtin, *Dialogic Imagination*, p. 84.
96. Bakhtin, *Dialogic Imagination*, pp. 87 f., 111 f.
97. Allen's translation, *Arabic Literary Heritage*, p. 57.
98. Kennedy, '*Maqāmāt* as a Nexus of Interests', p. 164.
99. This observation was in part inspired by Julia Bray's insightful response to my presentation, The Party of Abū al-Qāsim, at the 2010 meeting of the School of Abbasid Studies in Leuven, Belgium. It is also a subject addressed in Barletta, *Covert Gestures.*
100. Petronius, *Satyricon* 119–124: 295. Other examples include the *Canterbury Tales*, as well as the first book of Apuleius' *Metamorphoses*, in which travellers pass the time listening to a tale which, as it turns out, bears a close resemblance to the main plot.
101. Al-Azdī, *Sittenbild*, pp. vii–ix. Jaʿfar's command is more literally translated as 'Say some verses and make it short!'

102. *Ḥ*, 44, also translated in Chapter 1, p. 33.
103. Gabrieli, 'Sulla *Ḥikāyat*', p. 40. Translated by Ronan MacRory.
104. St. Germain, *Anomalous*, p. 155.
105. Montgomery, *al-Jāḥiẓ: In Praise of Books*, p. 117.
106. Mubārak, *Nathr*, p. 431.
107. See pp. 106, 112.
108. ʿAlī ibn Abī Ṭālib is discussed in Chapter 5, pp. 151–2. Al-Khaṭīb al-Baghdādī, *al-Taṭfīl*, p. 44.
109. Also see van Gelder, 'Jest and Earnest', pp. 84–5, which explains how humorous anecdotes were included in proportion to more serious fare in order to provide a break for the reader. He points out, however, that jest and earnest are not always easily distinguished from one another, and goes on to provide definitions for each term (p. 86 ff.)
110. *Ḥ*, p. 89.
111. Moreh, *Live Theatre*, p. 100.
112. Gowers, *Loaded Table*, p. 247.
113. Gowers, *Loaded Table*, pp. 246–7.
114. Hamori, *On the Art*, p. 72.
115. *Qaṣīda* poems and their traditional subjects of description are explained in the 'Translations' section of the introduction (pp. 6–7), where St. Germain's study of the *Ḥikāya* as a parody of these traditional poems is addressed.
116. Hamori, *On the Art*, p. 60.
117. The Barmakid family were highly influential in their roles as advisers to the Abbasid caliphs in the late-eighth to early-ninth centuries. Yaḥyā (d. 806) was the tutor of the caliph al-Mahdī (ruled 775–85), who looked up to him as a father figure. Yaḥyā's son, al-Faḍl, also served at the court of Harūn al-Rashīd. This poem can be taken to mean that even serious and important men such as the caliph and his advisers should make time for laughter.
118. From al-Masʿūdī, *Murūj*, cited in Nicholson, *A Literary History*, p. 260.
119. Mez describes several amusing examples of this. For example, the festival known as '*laylat al-maḥsūs*' (night of the touch) [sic], part of Lent celebrations in Baghdad, included cross-dressing, people riding backwards on wooden horses, and unrestrained socialisation between opposite sexes. A New Year's festival in Egypt included a mock market-inspector riding on a donkey, mud and water fights, and students throwing their teacher in a fountain (*Renaissance*, pp. 423–4).
120. Al-Nīsābūrī, a scholar from Khurāsān who also wrote about Qur'anic philology, is further discussed in Chapter 4, pp. 125–6, and Chapter 5, p. 137.

121. Nīsābūrī, *'Uqalā'*, p. 57.
122. Carroll, *Through the Looking-Glass*, p. 183.
123. *Ḥ*, p. 279.
124. Borges, *Collected Fictions*, p. 482.
125. See pp. 139, 158.
126. Jeanneret, *A Feast of Words*, p. 24. Montgomery notices a related paradox in the *khamriyya* genre of wine poetry ('Justified Sinner?', pp. 1–90).
127. Earlier versions of portions of this chapter were published in Selove, 'Who Invited?'

3

Crashing the Text

> Most statistical tables are parchingly dry in the reading, not so in the present case, however, where the reader is flooded with whole pipes, barrels, quarts, and gills of good gin and good cheer.
>
> Melville, *Moby-Dick*, 'The Decanter'

Since the *Ḥikāya* takes place at a party, we might expect to experience some of the pleasures of this party while we read the narrative. As Abū al-Qāsim compares Baghdad to Isfahan, copiously listing foods and goods in a tirade that occupies a very bulky portion of the narrative,[1] he seems to proceed in the order of events one would expect to experience if attending a feast. He begins with the overall atmosphere of the cities and their place-names, then describes their horses (on which, perhaps, we can imagine arriving at the party. He then describes clothing and houses (two of the first things we would notice upon arriving at a party). Some samples of these descriptions are found in Chapter 1.

Abū al-Qāsim concludes his description with wine, snacks, and entertainment, thus portraying, in his exhaustive and exhausting fashion, every luxury that could be presented at every stage of a party. He is by no means describing the party that he is attending in Isfahan; this party to which we are treated is a wholly hypothetical and encyclopedic feast of words. It feels highly specific, but, as stated in the previous chapter, only describes things not found in Isfahan. These are highly specific lists of things that do not, in the immediate context, exist.

Abū al-Qāsim's bombastic or grotesque poetry and prose stand in contrast both to his conversational exchanges with his half-stunned audience, as well as to the narrator's voice, which speaks rarely and simply. Thus the

Ḥikāya weaves contrasting registers of Arabic literature into a continuous dinner conversation. The more literary and elaborate the language, the less it seems to have to do with reality. To take an example (cited in the Chapter 1), Abū al-Qāsim, having hinted (not so subtly) that he is hungry, is asked what he would like to eat. He replies with an acrobatic recitation of literary requests not only for food, but for gifts fit for the guest of a king: a swift charger, a choice singing girl, fine clothing, and so on.[2] When the other guests complain that they are intimidated by these extravagant requests, Abū al-Qāsim makes a more realistic request in prose: 'A soft loaf, cheese so strong you weep, strips of local meat, tender and smiling, something from the ready foods of the market, and whatever lingering little bites you have around, like pickled snacks.'[3] Then the narrator describes food brought to the table even more simply, mentioning 'a platter of the cheese he had asked for, and some pickled things.' Though Abū al-Qāsim is not exactly cheerful at the spread, so simply described, he brightens up when a lamb dish follows, apparently forgetting his outrageous requests in verse, as well as his lists of elaborately superior items to which he implied the Baghdadis were accustomed. Thus the substance and intricacy of Abū al-Qāsim's requests for food are adversely proportional to the material sustenance given him to eat – only when he makes his request in simple terms is he fed. However, he greets each dish with another literary outburst.

Ibn Buṭlān (a Christian physician, d. 1066) composed, in addition to many learned works such as the famous *Taqwīm al-siḥḥa* (The Raising of Health, an influential text also in medieval Europe), a comic banquet text called *Daʿwat al-aṭibbāʾ* (The Physician's Dinner-Party). In this story, a stingy doctor invites a young aspiring physician to his house for dinner, but allows him to eat very little, describing how each dish could be bad for his health. Then the doctor's friends arrive at the dinner, and each discusses his medical field of specialty; the banquet consists of a multi-course meal of conversation, but very little actual food.[4] We can read this story, with its playful contrast of food with speech, as part of a Mediterranean banquet literature tradition. Ibn Buṭlān, himself a physician, was certainly influenced by the Greek tradition of medical and philosophical literature (citing Hippocrates and others in his comic banquet tale).

Classical banquet texts (Plato's *Symposium*,[5] the *Deipnosophistae*, and

Trimalchio's dinner-party being three of the most famous examples), explore the relationship of description and reality by setting various literary depictions of pleasure in a context of the real physical pleasure of a party. The introduction to the encyclopaedic portrayal of dinner conversation in the *Deipnosophistae* tells us, '[This] account is arranged to imitate the extravagance of the dinner party, and the book's structure reflects how the dinner was organized.'[6] As Davidson writes, 'Throughout the *Deipnosophistae,* there is a rather peculiar identification of words about food with food. There is an actual feast going on , but there is also a feasting on words, which is given much more emphasis',[7] later concluding, 'From this perspective Athenaeus' banquet becomes a banquet of Tantalus, thanks to a simple rule: you cannot talk properly with your mouth full. The feast of words is a feast of not eating, an anti-feast.'[8] Unusual patterns of consumption often play against unusual literary productions, or replace the consumption of food with the consumption of conversation and literature. The guests at the *Symposium* famously forego the typical pleasures of a banquet for the pleasures of an intellectual debate. In the Trimalchio's Dinner-Party, the shocking, colloquial speech of the social-climbing freedmen garnishes the indigestibly showy, hybrid foodstuffs provided by Trimalchio. Like the food, their speech is vulgar with 'a veneer of respectability and attainment'.[9]

The *Ḥikāya,* like the *Satyrica,* has been hailed as an early attempt at realism,[10] but, as in so many Mediterranean banquet texts, it replaces consumption itself with depictions of consumption, and conversation about absent food as well as sex, music, and material goods dominate the feast. This conversation does not match the food or events at the party itself in nature, as does the speech of the freedmen in Trimalchio's dinner-party. For although the dinner served at the Isfahani party seems modest enough, Abū al-Qāsim's elaborate foodstuff-laden speech leaves us with a feeling of dazzled overfullness. Thus the dinner conversation is not a representation of the food served, but a false display of a sumptuous feast, whose grand, luxurious offerings of gourmet meals, sex, cushions, and song – like Abū al-Qāsim's incredible ravings – are ultimately just indigestible words. Al-Azdī promises us a written mimesis of Baghdad, but throughout his text, there is a tension between words and reality, emphasised by this replacement of banquet with text. Abū al-Qāsim exchanges his wordy descriptions that may not describe something real in Baghdad, for real food at a party in Isfahan that is hardly described at all.

The frustration of being served a meal of words instead of real food is a theme found elsewhere in contemporaneous Arabic literature.[11] In al-Hamadhānī's '*Maqāma* of the *maḍīra*' a host, reminiscent of Trimalchio in his tastelessness, and almost as long-winded as Abū al-Qāsim, boasts so much to his guest about his food and household accoutrements that the guest becomes disgusted and unable to eat.[12] 'In [this story]', comments Daniel Beaumont, 'language does not disclose reality, it is rather a deafening roar which obscures it.'[13] Both al-Hamadhānī's '*Maqāma* of Famine' and '*Maqāma* of Fresh Butter' include teasing descriptions of delicious food maddeningly delivered to hungry people. In al-Khaṭīb al-Baghdādī's *al-Taṭfīl* (another eleventh-century Baghdadi party-crashing book), a man describes a pastry so deliciously that his friend grows angry. 'What do you think', says the man, 'O Abū al-ʿAbbās, about *jawzīnaj*'s light and flaky crust, and the power of its sweetness, drowning in sugar and oiled nuts?'[14] To which Abū al-ʿAbbās replies, 'If the pastry were only here right now, I would savour its presence so much more than your description. But as it is not, please let us do without more description as we must do without the pastry itself.'[15] In a similar anecdote, someone asks the famous gourmand and party-crasher Bunān what he thinks about the pastry *fālūdhaj*. 'Should anyone', he replies, 'asking about *fālūdhaj* in this earthly life refer to intellect or reason? You simpleton! Eat it!'[16] Other party-crashers in the same text advise avoiding conversation altogether during a feast, as it interferes with the chewing process, further highlighting the tension between talking and eating.[17]

On the other hand, almost all of these party-crashers get fed by means of their silver tongues; indeed in a society that values eloquence above all else, one may make a living eating one's words, so to speak, by exchanging praise and poetry for daily bread.[18] Party-crashers tend to earn their invitation to the party by facilitating the dinner conversation. In Xenophon's *Symposium*, to take a classical example, the conversation (that is, the text itself), would not have happened without the party-crasher Philippos. This joker arrives uninvited at a party at which the guests are dining 'in silence'. 'You all know that I am a jester', he announces, explaining that he has crashed the party 'thinking it more of a joke to come to your dinner uninvited than to come by invitation.' 'Well', replies the host, 'take a place; for the guests, though well fed, as you observe, on seriousness, are perhaps rather ill-supplied with laughter.'[19] Philippos' subsequent lame attempts at humour spur the previously silent guests on to lighthearted conversation. This conversation is the substance of Xenophon's *Symposium.*

Encolpius, the (anti-)hero of the *Satyrica*, is an outsider to Trimalchio's party, if not himself a crasher, and he asks his neighbour to explain the events and people that he sees there. His freedman interlocutor not only describes and explains the feast, but helps colour its seedy atmosphere with his low, colloquial language. Thus the parasite in banquet literature can serve in the role of intratextual narrator of the text as banquet. For example, the narration may occur through the perspective of an outsider to the feast: in Plato's *Symposium*, the narrator Apollodorus hears his tale from Aristodemus, who tagged along to the party with Socrates, uninvited by the host (Socrates makes witty literary excuses, saying 'To the feasts of the good the good unbidden go').[20]

In the *Ḥikāya*, the dominant voice is that of a party-crasher, Abū al-Qāsim. However, Abū al-Qāsim does not narrate the banquet that he sees, but rather drowns out its description and the comments of its Isfahani guests with his ravings. The *Ḥikāya* therefore resembles the tradition of narrating through the lens of an outsider to the feast, but with a twist, because at this party, the uninvited guest takes over the banquet, commanding as if he, the guest, were now the host. Likewise his speech dominates the text, shoving the scenery, the other characters, and reality itself to the margins. Thus, Abū al-Qāsim represents a twist on the typical role of the party-crasher as guide to the reader, because he makes the party his own.[21] The other guests at the party in the *Ḥikāya* at once represent the reality which is occluded by Abū al-Qāsim's grotesque speech – the simple Isfahani party about which we hear so little – and also the experience of the reader when confronted with the *Ḥikāya*, and with the representation of Baghdad as a literary bully, impossible to resist, difficult to question, eminently entertaining but nauseatingly overabundant.

In the few studies there have been of *Ḥikāyat Abī al-Qāsim*, a few key passages receive the most attention: the author's introduction, as well as the opening pages of the story in which Abū al-Qāsim himself is described, and various of the more outrageous and notable outbursts of the protagonist. But few notice the comments of the largely silent Isfahani party-goers, and the modest comforts offered at the banquet Abū al-Qāsim crashes. Indeed with uninterrupted quotations from the protagonist, some as long as ninety-three pages (in al-Shāljī's edition), it can be easy to forget that there is anyone else at the party. But by focusing on the few moments when Abū al-Qāsim converses with the other guests, one learns something about his character, and a little about the hypothetical but realistic party of people in awe of his shocking and literary speech, which is so overabundant that they hardly have time to digest what he is saying. Though the other guests attempt to engage him in conversation and laugh at his jokes, he never really allows them to respond as he drowns out their meal with his words.

The first to try to talk besides the protagonist is a 'tough guy' (*jald*) from the crowd, who puts an end to Abū al-Qāsim's initial display of piety. At first the crowd, moved by his passion, responds as the audience of a pious exhortation should – with pious tears. But after all, they are at a party, and soon the crasher is cut short with an abrupt 'Never mind!' from the tough guy, 'come off it, because everybody here drinks and fucks'.[22] This exclamation follows the tough guy's suddenly 'understanding' Abū al-Qāsim (*yafṭan lahu*), though what precisely he understands could be a subject for debate. Balda might argue that he 'understands' that Abū al-Qāsim is not really pious, as his tearful recitations suggest, but is only a rogue putting on an act (although, as previously stated in my introduction,[23] I do not agree with this reading, as I will detail in Chapter 5). The tough guy may understand, at least, that Abū al-Qāsim is not *merely* pious, or that he has more to offer in the way of unasked-for party entertainment than pious recitations. Either way, he suggests that Abū al-Qāsim's religious display is inappropriate for the party in question, composed of a group of people in a private convivial setting, who could handle sterner stuff in the way of entertainment.[24] Despite the tough guy's braggadocio, the crowd will soon realise that Abū al-Qāsim's conversation is in fact much more than they can handle.

The party-crasher smiles at the guest's remark, and changes his manner entirely until the end of the tale. He takes off his sombre hood (his *ṭaylasān*, 'a sort of official mantle of godliness' worn by religious scholars)[25] relaxes his demeanour, and begins to insult each member of the feast, asking the other guests first to identify one another, saying 'who is such-and-such?' We can compile a guest-list from their brief responses: first there is the host, then the 'tough guy', then an erudite man named ('for example') Abū Bishr, then a secretary (*kātib*), who is an 'important' (*khaṭīr*) man connected to the chief of a government bureau, his black servant, a man with a large turban, a visitor and friend of the leaders of the community,[26] a tanbur player,[27] a joker, a butler (*wakīl ṣāḥib al-dār*), a beardless youth, a man sitting silently, two friends, a lute player, a male and female singer, her two guards, a woman (not otherwise described), and a slave boy.[28] With the exception of the black servant, and the young and attractive members of the party (who are instead made the victims of Abū al-Qāsim's sexual aggression), each of these party-goers is lambasted by the crasher as being the worst of a stereotype: for example, the secretary is an illiterate idiot, an 'important' man in charge of guarding the duck guano on the river bank, and furthermore he only brings a servant to puff himself up. The guest introduced as a 'joker' receives the most scathing and lengthy round of insults, and almost leaves the party in humiliation.[29] This particularly aggressive act of the protagonist is telling, since Abū al-Qāsim, despite his entertaining behaviour, does not want to be identified as a joker himself. He may be funny, but he will not stand to be laughed at, and thus leaves the audience unable to interact with him at all. He denies that every other guest properly suits his category or profession, and himself refuses to be classified in a category.

Any audience in a medieval or Mediterranean banquet setting would recognise a fool as a familiar dinner guest, his function including humorous insults for each guest. However, Abū al-Qāsim refuses to inhabit this paradigm, and during his insults the crowd seems uncertain whether they should try to defend the other guests or to play along, trying both strategies alternately. At first, they are hopeful that he is there to entertain them, and that he is himself another joker: for example, they join him in a hearty chorus of 'his beard in your ass!'[30] (Abū al-Qāsim's favourite insult, directed in this case toward the house steward). Thus, they encourage his verbal barbs.[31] Indeed, the crasher soon demands to be fed, reinforcing the impression that he is engaged in an exchange. So the crowd continues to play along when he exclaims, 'There's not a living man among you, Arab or otherwise!' 'What do we say?' they ask, encouraging him to go on, 'What do we do?' Abū al-Qāsim responds with sexual and financial advice so immoral that one guest makes the terrible mistake of laughing out loud; the protagonist, who hates above all to be laughed at, erupts into a blasphemous and scatological tirade whose fury he hardly matches again during an entire day of scatological outbursts.[32] 'What are you laughing at?' he asks the guest, adding that the sound alone hurts his ears. As Balda puts it, Abū al-Qāsim is *hors la loi* of discourse, impossible to respond to or even laugh at.[33]

'You've gone too far!' the guests exclaim for the first time during Abū al-Qāsim's vicious tirade against the 'joker'. And during his excoriation of their hometown, Isfahan, the guests again complain that he is 'going too far', exceeding the bounds of proper conversation even for a jester. Later Abū al-Qāsim bridles violently when they ask him for some entertaining '*ḥikāyāt*', exclaiming, 'Do you think I'm a buffoon?' For though he resembles nothing so much as a entertainer who sings for his supper,[34] only by addressing him in a tone of meticulous respect do the guests manage to guide his conversation to several topics that interest them.

Abū al-Qāsim not only refuses to allow the guests to respond to him in an ordinary way, he also refuses to let them remain silent, for the last man he attacks is a guest sitting silently through his verbal rampage.[35] This attack is finally too much for the other party-goers, for before the crasher can really get started on his new round of insults, one guest remarks, 'I've had enough of Abū al-Qāsim and his conversation … The weather today is lovely, and the sky is clear. Let us drink three toasts to this beautiful blue sky!'[36] Ever the enemy of anything resembling normal human conversation, Abū al-Qāsim bridles at the suggestion that they talk about the weather. In fact, this offending reference to the Isfahani weather sets off the theme for the majority of his conversation: the comparison of Isfahan with Baghdad. The Isfahani crowd listens uncomfortably, occasionally encouraging him with a polite but curt 'speak, Abū al-Qāsim', occasionally complaining again that he has gone too far. After a very long rant, the crasher demands once again to be fed. Having satisfied his appetite, he reverses his former positions, now blaming Baghdad and praising Isfahan. The crowd is baffled by his fickle use of words, and though he embarked on his comparison in angry response to what he considered a trivial, meaningless use of language (the comment about Isfahani weather, which he characterises as 'completely meaningless, like splashing water about'),[37] his own words seem equally substanceless in their way, full of tropes and clichés he is willing to use for either city.

The other guests have tried and failed to interact verbally and socially with their monstrous fellow diner. One may well ask why they put up with him at all. More than one answer is possible, but like much that is obscene or frivolous in contemporaneous Arabic literature, both al-Azdī and his character Abū al-Qāsim win themselves the partial indulgence of their audience by promising to teach them something. Al-Nafzāwī, for example, author of the famous fourteenth-century sex manual *The Perfumed Garden*, promises in his introduction to teach his patron and his readers about sex.[38] Nor is his work without educational content, though what really hooks us, it seems, is its page-turning obscenity. Al-Azdī and Abū al-Qāsim also promise to teach us something, but, although they are not as uniformly obscene as a sex manual, they are shocking enough that the modern reader, and probably the Isfahani audience, may endure some annoyances just to hear what they will say next.

Al-Azdī promises a portrait of Baghdad that has proved enticing to modern scholars of the city, some of whom seem to endure the obscene content only to benefit from these educational insights into Baghdad. Likewise, in the story, Abū al-Qāsim renews the attention of an offended audience by offering to teach them how to be 'gentlemen' (*nās*). Both enticements of educational content are immediately undercut, however: al-Azdī, in the same introduction in which he promises a portrait of Baghdad, as seen above, characterises his work as a mere 'evening tale' (*samar*), thereby designating it as a kind of story with very low truth value. And because Abū al-Qāsim provides such bad advice, a kind of 'modest proposal' to plunge into debt and 'fuck [anything that moves]', it would be difficult to consider his speech anything but a mockery of the didactic. It is true that we can turn to different sections of the *Ḥikāya* for poems about horses, fruits, beautiful and ugly people, and many other topics, and that it can serve, therefore, as a kind of encyclopedia of *adab*, a word that means both literature and good breeding (as a familiarity with literature was required of a participant in polite society). But the immediate source of this *adab*, Abū al-Qāsim, cannot exactly be considered a guru of high culture – although he knows his literature, his manners are appalling, and he smells bad and drinks excessively.

Given the type of playful conversation expected at a banquet, any text set at a party invites scepticism concerning the sincerity and reliability of the often drunken dinner-conversation it depicts. We cannot trust even the *Deipnosophistae*, which depicts learned diners quoting copiously and obscurely, and which is therefore valued by modern scholars as a source of fragments otherwise lost from Greek literature. Indeed, it is considered by many as a dry encyclopedic collection of trivia. But when the learned diners themselves cast doubt on the reliability of their fellow-diners' scholarship, when pedantic displays of learning are openly mocked, and when certain names (among a sea of obscure words and names) are clearly invented based on literary analogies, then the author/compiler seems to play with the reader as often as he instructs her. We cannot always know which is which, and that not-knowing is part of the game.[39]

As in the *Ḥikāya* and its forebears, it is difficult in many Arabic works of *adab* to distinguish between the sincerely didactic and the playful and parasitical.[40] Al-Khaṭīb al-Baghdādī's *Kitāb al-taṭfīl* (*Art of Party-Crashing*) for example, presents prophetic hadith on hospitality and table manners side by side with party-crashing advice on how to avoid table conversation in order to facilitate rapid chewing. It presents musings from the first successor to the prophet Muḥammad side by side with misuses of Qur'anic quotations by party-crashers angling to be fed. I do not suspect the piety of the author (although according to some he was expelled from Damascus as punishment for dalliance with a handsome boy, and possible drunkenness).[41] These allegations aside, he was a famous and dedicated scholar of prophetic hadith, and he begins his text with citations and actions of the prophet of Islam. If his work has a morally didactic purpose, it must be to justify the party-crasher's actions, as any host who would deny them entry is a miser, whose behaviour is at odds with the generous behaviour of the prophet Muḥammad himself. As for the crashers, however, is their advice to 'not fraternise', nor 'pity the weakness' of an aged guest when elbowing in for meat, meant to be taken seriously?[42] Rather, we should laugh at their single-minded greed. But the compiler makes little effort to distinguish these anecdotes from the simple advice on table manners and etiquette apparently issued in all sincerity.[43] Some pieces of advice lie between earnest and jest: one anecdote warns that reticence and abstemiousness in a guest may hurt the host's feelings;[44] no doubt they may, but quoted, as it is, by a party-crasher, can we see through to his intentions to use this sound advice as a thin excuse for eating as much as possible at someone else's expense?

Abū al-Qāsim's speech resembles some of these anecdotes by hovering between the genuinely informative and the laughably outrageous. Sometimes his conversation even resembles the deliberately nonsensical speech mocked in *al-Taṭfīl*, for example. In *al-Taṭfīl*, Bunān, the famous party-crasher, warns against going to dinner with members of certain professions (such as tailors), and goes on to mock their gossipy, babbling speech ('I'll cut a robe for a third and a dirham of a third, and two dirhams of a third, then three for a half and two dirhams with a half and three dirhams!')[45] Abū al-Qāsim sometimes resembles Bunān in this anecdote, mocking professions with his stereotyping insults. But sometimes he almost resembles the babbling objects of Bunān's mockery.

It is, indeed, difficult to verify that some of what he is saying might not be made-up nonsense words, or something like the talk of the rambling doctors, barbers, or tradesmen who speak learned nonsense at aggravating length in other Arabic texts.[46] For example, in a story by al-Tawḥīdī (thought by some to be the author of the *Ḥikāya*), a sick man is plagued by a learned man who babbles at his bedside with utterly nonsensical advice ('It falls to me in what does not fall to one other than me or like me in one who could as well have been me or as though he were of my age or was known by what is not known of him to me that I see that you are not keeping anything but a diet above what is necessary and below what is not necessary', he begins, and goes on from there).[47] These rants, filled with learned-sounding language in grammatically logical but meaningless bundles, betray the author's disgust with specialised discourses, or with those who affect learning and linguistic skill but actually babble incomprehensibly. In fact, the first guest to receive Abū al-Qāsim's scathing verbal attention is introduced as 'a man of exceeding erudite knowledge' (*rajul fāḍil adīb*) and Abū al-Qāsim attacks this man with the kind of nonsense speech described above, asking 'Doesn't the shaykh not understand how he is not knowing?',[48] after suggesting that he 'read *The Retardation of Knowledge*, or *The Book of Forgetting Stuff*, and studied at the Institute for the Deficiency of Understanding'.[49] He also attacks the man's name, Abū Bishr (loosely, 'Mr Happy') as nonsensical, complaining that to call such a sorry specimen 'Mr Happy' is like calling a sewer pipe 'Mr

Clean'.[50] And inappropriate naming is perhaps the most concrete example of the often arbitrary or misleading nature of the relationship of language to reality.

Though Abū al-Qāsim does not always rave in an altogether nonsensical fashion, he raves nevertheless, enough to put the didactic value of his speech very much in question. Given the authority of the Arabic language, the locus of eloquence and truth embodied in the Arabic Qur'an (itself an often incomprehensible book),[51] these portrayals are generally subversive, and evoke anxieties about who has knowledge in a world of duelling discourses; the hadith scholars, the debaters, the poets, and the street preachers all make warring claims to power based on various manipulations of the Arabic language and the authority that it represents.

Abū al-Qāsim's esoteric speech also can resemble expositions of beggar's slang, as in an anecdote in al-Jāḥiẓ's *Kitāb al-bukhalāʾ* (Book of Misers), in which a miser gives a beggar a more valuable coin than he intends, and then takes it back. When the beggar complains, the miser says that he can tell the beggar is not worth the valuable coin, because he is so well versed in different types of beggars. He then awes his interlocutor into silence with a jargon-filled description of all the types of beggars and tricksters found in his society, each of which he provides with a label and a definition. He is thus able to cover his unsavoury act of miserliness with his arcane linguistic knowledge.[52] This story most closely resembles a passage in the *Ḥikāya* in which the other guests ask Abū al-Qāsim if he knows any sailor's expressions. The crasher boldly obliges with a long list of obscure words such as *māshūka*, *kanūr*, *kadl*, and *mahār* (about which words al-Shāljī can only confess, 'I didn't understand them').[53] Abū al-Qāsim's verbal display is impressive, but without any definitions for his terms, his speech provides not even the didactic benefits of that of al-Jāḥiẓ's despicable miser.[54]

Party-crashers themselves crash both physical banquets and banquets of words like the Qur'an,[55] using these discourses parasitically, and twisting them to their own purposes. *Al-Taṭfīl* tells of a party-crasher who used various quotations from the Qur'an with numbers in them to impress his host into feeding him that number of delicious pastries. Another crasher takes a Qur'anic verse out of context to reassure the host that he has come not to steal his women, but his food.[56] Abū al-Qāsim may be no less an opportunistic user of the Arabic language. He is certainly a master of all discourses, and uses and abuses them at whim, just as al-Azdī openly admits to passing many different works off as his own. Adam Mez begins his introduction to his edition by saying, 'If someone posed the embarrassing question to our author, [of what would happen] if someone told his sentences, 'Go back to where you came from', then very little worthwhile (though a lot disreputable) would remain.'[57]

As a sympotic text, the *Ḥikāya* is hardly unique in that regard, again resembling, for example, the quotation-ridden *Deipnosophistae*. As Davidson writes, 'The "real" banquet, described as taking place during the conversations of the *Deipnosophistae*, is evoked in the learned discussions which are provoked by it, but the author spends only the tiniest fraction of his text in describing it.' This banquet, held in Rome, of Greek literature, seems a 'separate world', whose sometimes tiresome profusion of quotations exists within their own system of order, and hold the questions of 'life, death, and representation'.[58] But the conversation of Athenaeus' learned banqueters seems sober and reasonable compared to the near-monologue that takes place during the banquet in Isfahan, with all its drunken outbursts and disorienting atmosphere. The next chapter will further address language that is difficult to understand, nonsense language, and language that at first seems to offer us information, but which we at last find elusive in meaning.

Notes

1. This section, as I describe it here, begins on *Ḥ*, p. 91. The description of entertainment begins on p. 187 and can be considered to continue to p. 270, where he returns to nostalgic memories of Baghdad and then finally demands to be fed (p. 274).
2. Translated in Chapter 1, p. 48.
3. *Ḥ*, p. 277.
4. Compare to Apuleius 1.26: 505–24, in which the protagonist's miserly host forces him to talk all night without offering him anything to eat. The unhappy guest then goes to bed having dined only on words (*cenatus solis fabulis*). Also see 'The physician's dinner party', in Kennedy 'The *Maqāmāt* as a nexus of interests', pp. 171–75.
5. Dimitri Gutas has shown that even Plato's *Symposium* was not unknown to Arabic writers. Gutas, 'Plato's *Symposion*', pp. 36–60.

6. Translated by Paola Ceccarelli.
7. Davidson, 'Pleasure and Pedantry', p. 297.
8. Davidson, 'Pleasure and Pedantry', p. 303.
9. Schmeling, *Commentary*, p. xxx.
10. Erich Auerbach credits Petronius with producing a representation of a social milieu by means of mimetic speech, *Mimesis*, p. 30.
11. This topic is addressed throughout van Gelder, *Dishes*, as well as in Gowers, *Loaded Table*, which explores similar tropes in Latin literature.
12. Al-Hamadhānī, *Maqāmāt*, pp. 104–17.
13. Beaumont, 'Mighty and Never Ending', p. 159.
14. *Jawzīnaj* is a kind of baklava made with walnuts.
15. Al-Khaṭīb al-Baghdādī, *al-Taṭfīl*, p. 105.
16. Al-Khaṭīb al-Baghdādī, *al-Taṭfīl*, p. 155.
17. Al-Khaṭīb al-Baghdādī, *al-Taṭfīl*, pp. 111, 131.
18. The exchange of words for a living was practiced by more than just party-crashers, of course. Devin Stewart argues in his article, 'Professional Literary Mendicacy', that the trickster/beggar hero of the *maqāmāt* may be inspired by the life of secretaries and men of letters who sold their literary skills to make a living.
19. Xenophon, *Symposium* 1.11–14.
20. Plato, *Symposium*, 174B. This is a sentiment the prophet Muḥammad himself would have approved, as demonstrated by early chapters of al-Khaṭīb al-Baghdādī, *al-Taṭfīl*, in which the prophet brings uninvited guests along to parties, or lets strangers follow him on his way to a feast.
21. This in itself is not unusual behaviour for Arabic party-crashers, who seem to question the very premise of ownership by their behaviour. In one story in *al-Taṭfīl*, for example, a very pushy crasher wonders silently to himself, 'Whose house is this?' and then answers his own question saying, 'It's yours, man, until someone says otherwise' (al-Khaṭīb al-Baghdādī, p. 112).
22. *Ḥ*, p. 55.
23. Balda, 'Marginalité', p. 383. See above, p. 17.
24. For a full description of such gatherings and the types of speech one might have heard at them, see Ali, *Arabic Literary Salons.*
25. Kennedy, '*Maqāmāt* as a Nexus of Interests', p. 165.
26. Abū al-Qāsim describes this man as a party-crasher in an insulting tone that is definitely hypocritical, given that he himself is crashing a party.
27. A tanbur is a long-necked stringed instrument.

28. For more details on the guest list, please see the painting acknowledgements at the beginning of this book.
29. *Ḥ*, p. 13.
30. *Ḥ*, p. 80.
31. The host even complains, 'There's no one left in the party for you to mention but me!' (*Ḥ*, p. 80) inviting a rain of insults on his own head (though these are relatively restrained; Abū al-Qāsim does seem reluctant to bite the hand that might feed him). The host's complaint about being left out of the fun suggests that Abū al-Qāsim's insults could still serve the socially recognised function of entertaining the guests, and exchanging, in a semi-professional manner, words for food; this kind of insult could be amusing and even flattering. The emperor Vespasian is said to have similarly solicited a joker's ridicule, and Carlin Barton explains in her *Sorrows of the Ancient Romans* how insulting the dinner guests may ward off envy and evil eye (p. 108 ff.). In her *Garden of Priapus*, Richlin likens the satirist to the garden god Priapus, whose phallus threatens and wards off the 'chthonic forces' (p. 113).
32. See Chapter 1, pp. 39–40 for a translation of this passage.
33. Balda, 'Marginalité', p. 392, as mentioned in the 'Scholarship' section of the introduction, p. 17.
34. St. Germain sees him as exactly that in *Anomalous*, p. 117.
35. *Ḥ*, p. 88.
36. *Ḥ*, p. 89.
37. *Ḥ*, p. 90.
38. He introduces his work as fulfilling a patron's request for a guide to things stimulating intercourse and things that hinder it, as well as methods to enlarge the penis and other similarly useful information (al-Nafzāwī, *Rawḍ*, p. 25).
39. These playful features of Athenaeus' work were described in detail by Ceccarelli, 'Athenaeus: A Walking Library?' She focused especially on the following passages: 13.610cd ('Learning, much learning – than which there is nothing more empty!'), and 14.644a–648c (on different types of cakes).
40. Connors shows how the *Satyricon* also regularly uses morally didactic poetry in contexts that undercuts its meaning. She also provides a valuable exploration of parodic use of poetic embellishment and quotation, as well as issues of representation and power (*Petronius the Poet*, passim).
41. Malti-Douglas, 'Controversy', pp. 115–31.
42. Al-Khaṭīb al-Baghdādī, *al-Taṭfīl*, p. 131.
43. See, for example, al-Khaṭīb al-Baghdādī, *al-Taṭfīl*, p. 144.

44. Al-Khaṭīb al-Baghdādī, *al-Taṭfīl*, p. 144.
45. Al-Khaṭīb al-Baghdādī, *al-Taṭfīl*, p. 147.
46. See van Gelder's 'Amphigory' (mentioned further below, pp. 118n, 126). It is difficult to verify because all of Abū al-Qāsim's words cannot be defined. Most of them can be, however. For some translated examples of true Arabic nonsense, see Al-Hamadhānī, *Maqāmāt*, trans. Pendergrast, pp. 131–4, or the tale of the mad barber in Haddawy (trans.), *The Arabian Nights*, p. 254 ff.
47. Yāqūt, *Udabā'* V: 1923.
48. *Ḥ*, p. 57, trans. van Gelder.
49. *Ḥ*, p. 56. This section is translated in Chapter 1, p. 36.
50. *Ḥ*, p. 57.
51. Interpreters of the Qur'an recognise certain passages as ambiguous (*mutashābihāt*), whose meaning can only be fully comprehended by God (see Kinberg, 'Ambiguous').
52. Al-Jāḥiẓ, *Bukhalā'*, pp. 46–53.
53. *Ḥ*, p. 318. The short vowels in these words are simply guesses.
54. Al-Jāḥiẓ, *Bukhalā'*. Mubārak addresses this passage and the similar passage on swimming strokes, comparing them to Abū Dulaf's *Qaṣīda sāsāniyya*, a poem of beggar's slang (*Nathr*, pp. 419–21, 432), translated and discussed by Bosworth in his *Medieval Islamic Underworld*. Van Gelder notes in his 'Amphigory', 'Since the Arabic language is replete with rare and obscure words, great erudition is required to avoid being fooled at times by words with a false air of authenticity', p. 16. He quotes a poem by Bashshār ibn Burd (d. c. 167/784), written from the perspective of a donkey to another donkey, that mocks the tendency of some writers to use obscure language: 'She has a smooth cheek | like a *shayqurān*'s cheek'. When someone asks him for the meaning of *shayqurān*, he replies: 'It is one of the recondite words of the donkeys, so if you come across one, ask him'.
55. The Qur'an has indeed been referred to as 'God's Banquet', reputedly by the prophet Muḥammad (see van Gelder, *Dishes*, also published under the title *God's Banquet*, pp. 39–40).
56. Al-Khaṭīb al-Baghdādī, *al-Taṭfīl*, pp. 105–6, 116–17.
57. Al-Azdī, *Sittenbild*, p. v.
58. Davidson, 'Pleasure and Pedantry', p. 303.

4

Mujūn is a Crazy Game

'De balena vero sufficit, si rex habeat caput, et regina caudam.' Bracton 1.3, c. 3 ... A division, which in the whale, is much like halving an apple; there is no intermediate remainder.

Melville, *Moby-Dick*, 'Heads or Tails'

'"Come!" cried old Omar. "Let us drink, and break into new patterns the tedious roof of heaven!"'

F. V. Morley, *My One Contribution to Chess*

Despite the topsy-turvy atmosphere of the text, the *Ḥikāya* may invite readers to use it as an encyclopedia of material goods or otherwise as a source of historical and antiquarian interest for a number of reasons.[1] One of these reasons is its obscenity, or focus on 'low' topics not typically addressed in other forms of literature. The *Ḥikāya* often employs a literary style known as *mujūn*, distinguished by its focus on low topics and use of obscene vocabulary. In 'Arabic *Mujūn* Poetry', Julie Meisami writes: 'Discussions by Arab scholars typically adopt historical, sociological, or biographical approaches' to *mujūn* literature, whereas 'the literary aspects of *mujūn* poetry have received little attention.'[2] We must strive to understand the literary definition of *mujūn* itself, however, before attempting historical or antiquarian readings of the *Ḥikāya*. This is not easily done, and indeed Zoltan Szombathy devotes an introductory section and an appendix of his book on *mujūn* to the task of defining this word alone.[3]

Here, in discussing the chess game played in the *Ḥikāya*, I will show that *mujūn* itself can be thought of as a kind of game, and that thus the style of language known as *mujūn* is by definition playful. In the introduction to his edition of the text, Mez identifies the chess game as the only section made of entirely original material in the *Ḥikāya*.[4] Here I will argue that the chess game can be read as a synecdoche for the work as a whole. With its specific and familiar-sounding details, it tempts us to try to reconstruct the game exactly as described. Upon closer examination, however, we find that the events described are impossible to reconstruct because they are each presented only as hypothetical examples of events, as well as being couched in playfully disorienting language. But before exploring the chess game, I will begin with a further discussion of *mujūn* by examining a story from al-Zubayr ibn Bakkār's ninth-century *al-Akhbār al-muwaffaqiyyāt* (Reports Compiled for al-Muwaffaq [d. 891]), the regent of the caliph al-Muʿtamid. This story describes not a type of literature, but a type of behaviour defined as *mujūn*.

This story is about four sons who are accused before the caliph al-Maʾmūn of unseemly *mujūn*.[5] When the caliph confirms the accusation by discovering the silly statements engraved on their signet rings, one son explains that they learned their silly behaviour from their father, who is the treasurer in al-Maʾmūn's court. By way of example he tells a story of the long and harrowing night in which his father discovered that there was no more sugar in the house. In this story, his father angrily demands that his entire household stand on one foot until the servants can purchase more sugar in the middle of the night. While his tired family abides by his arbitrary command, he raves on a wide range of subjects such as property taxes, the improper behaviour of women, and (when the servants finally do return with sugar) the minute details of lawful Islamic buying and selling practices.

Since this story is presented as evidence of their father's *mujūn*, we can look to it for information on the defining characteristics of *mujūn* itself. ʿAlī ibn Ṣāliḥ, the treasurer, in his night-long act of *mujūn*, establishes an alternative set of rules for his party to live by (that everyone must stand on one foot until he receives sugar). Meanwhile, as noted by Stefan Leder in his 'Prosa-Dichtung in der aḫbār Überlieferung', his absurd behaviour undercuts any wisdom that may be found in his discussion of the laws that regulate daily life (like property taxes and buying and selling practices). As for ʿAlī ibn Ṣāliḥ's speech, Leder writes,

> It shows that he considers the lack of sugar a disaster – itself already a laughable exaggeration of a really inconsequential problem – and that he

> goes from the particular to the general to the point that he seeks to integrate his subjective perception into the level of general truth – the truth of the revelation and the sunna.[6]

Mujūn, therefore, can describe behaviour that follows a set of bizarre rules, whose enactment parodically echoes the logic of the daily laws under which we live (though this may sometimes have the effect of confirming the logic of these laws).[7]

In this sense, *mujūn* is like playing a game. Game-playing also demands that we abide by a set of arbitrary rules. Paradoxically, abiding by these rules constitutes a break from abiding by the rules of daily life. So to relax and have fun, instead of temporarily eschewing laws altogether, we sometimes prefer to adhere to a very strict regimen of separate rules. Leslie Kurke's *Coins, Bodies, Games, and Gold* shows that games as portrayed in certain Greek texts have a similarly contradictory status as at once mirroring the laws of life outside the game and representing a distraction from that life.[8] At the same time, however, Kurke describes how game logic and life are directly related; for example, the game board is likened to a city, and by implication, life in a city is likened to a game, as, for example, in fragments of Herakleitos.[9] One such fragment ('Time is a boy playing, playing *pessoi*; kingship belongs to the boy') portrays certain games, like the strategic board game *pessoi*, as reflecting the order of the cosmos. As Charles Kahn interprets it,

> … these moves follow a definite rule, so that after one side plays it is the other's turn, and after the victory is reached the play must start over from the beginning. The rules of the *pessoi*-game thus imitate the alternating measures of cosmic fire.[10]

As shown below,[11] later texts, and some medieval Arabic texts, make similar analogies, especially with regard to the game of chess.

Of all the games Abū al-Qāsim could have played, chess is certainly not the most absurd, or the one most readily likened to *mujūn*. Abū al-Qāsim in fact proposes that they play either chess or backgammon (Abū al-Jalab and Abū al-Ṣannāj, his nonsense jargon nicknames for the games).[12] Backgammon, according to medieval game lore, was thought to symbolise a more fate-driven universe beyond our control, while chess symbolised the logical, careful choices of an agent with free will.[13] Indeed in his humorous work, *My One Contribution to Chess*, F. V. Morley complains that by careful study of the patterns of chess, the game may cease to be a game altogether, and become something of a mathematical exercise, precluding a sporting match between two players of unequal experience or knowledge of, for example, opening moves.

Morley sees the game of chess as a 'dromenon', defined as 'a pattern of dynamic expression in which the performers express something larger than themselves …'[14] In other words, chess is a model for life, and is often considered, like Abū al-Qāsim, to be a microcosm with certain heuristic utilities. When working in a government office dedicated to the resolution of labour disputes (the War Labour Board), Morley found relief from the mixture of chaos and red tape that characterised his job in the relatively orderly representation of the universe provided by the game of chess.[15] Having lived through two world wars, he even recommends chess as an activity over which families may reconnect with returning veterans. Chess, long thought of as a representation of war, may provide a more orderly and socially pleasurable representation of real life which, though also played according to rules, can nevertheless seem senseless in comparison.

Chess is similarly considered an orderly representation of the universe in Arabic sources contemporary with the *Ḥikāya*:

> *Look, and you will see chess revolving like fate*
> *Day and night, misfortune and blessings.*
> *Its Mover remains, all the rest of it passes,*
> *And after annihilation, it is revived and its bones resurrected.*[16]

Kilito discusses an effect found in the *Maqāmāt* of al-Ḥarīrī, in which each episode contains in miniature the overall themes found in the work. These episodes, writes Kilito, portray history as cyclically recurring, until it is abruptly brought to an end.[17] If the *Ḥikāya* is understood as only one such episode of a cycle in extreme close-up (a possibility discussed in 'Those Camels have Passed'), then we can liken it to a single chess game among a string of chess games, at once containing the essentially unreal and open-ended potentialities of a game, and providing an ordered representation of the cosmos.

In his *Gambling in Islam*, Franz Rosenthal summarises fans' and critics' reactions to chess by saying it was praised by some for its 'usefulness for improving the mind and teaching military strategy', but on the other hand, it was criticised as 'too engrossing and causing neglect of [...] religious duties'.[18] So, despite its therapeutic orderliness and status as a dromenon, chess remained a distraction from real life. In an (almost certainly apocryphal)[19] condemnation of the game, the Prophet's cousin and son-in-law ʿAlī ibn Abī Ṭālib says, 'Chess-players are the biggest liars of all or among the biggest liars, saying "I killed", when they did not kill' (a reference to the term 'check mate', or *shāh māt,* which means, 'the king died').[20] This condemnation, like the narration of the chess game in the *Ḥikāya,* links game-playing to a type of speech with a complex relationship to the truth.

A *maqāma* appearing in Ibn Abī Ḥajala's fourteenth-century *Unmūdhaj al-qitāl fī naql al-ʿawāl,* (The Model of Fighting: On Grandmasters' Moves)[21] begins with the protagonist entering a town 'on whose board the pawns of the stars are drawn up in a line', and then proceeding to 'go round in it like the rook on the board'.[22] This story goes on to tell the story of a trickster who, much like Abū al-Qāsim, has a witty remark for every chess move made, thus linking the game to a playful kind of speech. However, Ibn Abī Ḥajala's book on chess includes a description of the games of chess and backgammon as representations of different models of the universe. As mentioned above, backgammon represents a deterministic universe, whereas chess represents a universe in which an agent has free will.[23] The specific numbers of the pieces used in these games and the manners in which they move are said to reflect specific features of the universe, such as the number of months in the year and the movements of the planets.[24]

These description of game-as-microcosm closely resembles descriptions of man-as-microcosm found, for example, in the thirty-fourth epistle of the secret philosophical society known as the Brothers of Purity, or Ikhwān al-Safāʾ (tenth/eleventh century). In this epistle, 'Fī maʿnā qawl al-ḥukamāʾ anna al-ʿālam insān kabīr' (About the meaning of the wise men's saying that the world is a giant human),[25] the Ikhwān describe how the course of the planets, the different types of human personalities, and even the number of the days of the week are all reflected in the various organs and characteristics of individual human beings.[26] Like chess, man (as well as the *Ḥikāya* itself) is represented as an intersection of disparate parts that together form a cosmic whole. But when these representations of the universe are presented as nonsensical or divorced from reality, they may cast doubt on man's perception of the universe itself, highlighting our capacity to live under nonsensical rules or to be convinced by eloquent and seductive but perhaps ultimately misleading speech.

In the poem cited above, the reordering of the chess board after a game is imagined as a metaphor for the day of resurrection. At the end of the *Ḥikāya*, after Abū al-Qāsim has passed out in drunkenness along with his exhausted fellow guests, he wakes up in the morning and resumes his pious manner and appearance exactly as at the beginning of the book, as if none of his raving or drunken insults had ever been enunciated. Rather than providing a comfortingly orderly representation of the universe, however, this resurrection seems to cast doubt on the reality of anything – of his pious demeanour when entering or exiting the party, or of the whirlwind of verbosity that occupies the bulky middle section of the text.

The *Ḥikāya*, though representing the passing of time with almost unprecedented realism, inhabits a topsy-turvy time in which day and night themselves are reversed and confused. As mentioned in Chapter 2, when Abū al-Qāsim and his fellow guests begin their game of chess, Abū al-Qāsim recites a nonsense poem expressive of this sense of topsy-turvy time:

> He would begin by advancing his pawns, and reciting some nonsense (*hadhayān*) by way of opening the game:
>
> > *We went out early, at dawn, in the night, in the evening, after midday,*
> > *And hunted rabbits, jackals, and wolves, but donkeys got away!*[27]

In *Nonsense: Aspects of Intertextuality in Folklore and Literature*, Susan Stewart offers a list of five ways to fail to make sense. Though she emphasises the fluidity and incompleteness of this list in her preface, we can nevertheless easily find the category 'discourse that denies itself', which best suits Abū al-Qāsim's poem.[28] As an example of this category of nonsense, she quotes a poem that begins: 'Ladies and jellyspoons: | I come before you | To stand behind you | To tell you something | I know nothing about. | Last Thursday | Which was Good Friday | There will be a mothers' meeting | For Fathers only . . .' Both Abū al-Qāsim's poem and this poem sound as if they refer to a specific time with detailed language ('early late at night after the start of day' and 'Last Thursday | Which was Good Friday') but each added detail actually detracts from our information by rendering the date described impossible and contradictory. Stewart begins her description of 'discourse that denies itself' by saying:

> The metacommunication necessary for the message 'This is play' or 'This is a fiction' implicitly carries a denial and a criticism – a denial because of the status of the representation as an activity that is framed as both real and non real, and a criticism because the discourse has been framed, set off, and is examinable from many sides and able to be manipulated.[29]

Thus this type of nonsense in the *Ḥikāya* further contributes to the tension between literary representation and reality felt throughout the text. Stewart's nonsense category of 'discourse that denies itself' best suits not only Abū al-Qāsim's speech during the chess game, but nonsense in the *Ḥikāya* more generally.

But as for the type of speech explicitly designated as nonsense or raving in the chess game, aside from the familiar-sounding example quoted above, we also have the following: 'Umm Razīn, she shat in the bread, | "It helps it rise!" Umm Razīn said.'[30] Here the narrator writes that Abū al-Qāsim raves deliriously (*yahẓī*), which may suggest that he is deranged or mad. But what type of madman is he? Al-Nīsābūrī's *ʿUqalāʾ al-majānīn* (Wise Fools)[31] provides a catalogue of different types of madmen,[32] for example, those driven crazy by jinn or demons (*mamsūs*), or those driven mad by love (*al-ʿāshiq*). None of these seem to suit Abū al-Qāsim. Al-Nīsābūrī also lists, however, a number of types of people who only seem to be mad or who pretend to be mad, in order, for example, to entertain others for money or food (like fools and jesters at a banquet), or those who pretend to be mad just to while away the time. These categories could be argued to include Abū al-Qāsim; furthermore, to pretend to be mad to while away the time is close to our definition of *mujūn* as a type of outlandish behaviour or speech practiced by way of a game. Those categories, however, who are most interesting to al-Nīsābūrī, and, as argued below,[33] probably best suit Abū al-Qāsim, are those categories of people who pretend to be mad in order to conceal that they are beloved of God, or who only seem mad to people so wrapped up in the mundane affairs of this earthly life that they are themselves, unbeknownst to them, as good as crazy. The ravings of the 'wise fools', those beloved of God, who only seem to be mad because of their disregard for the rules of everyday life (and these occcupy the bulk of al-Nīsābūrī's book), often contain nuggets of divine wisdom concealed in their gibberish. Some of Abū al-Qāsim's speech might fit this description as well,[34] but this could hardly be argued for the dirty little doggerel about Umm Razīn, which better suits a 'madman' entertainer, or one who raves to while away the time. We must conclude that Abū al-Qāsim is multiple different types of madmen at different points in the story; indeed, as a human microcosm, we can suppose that he is potentially all of these types of madmen at once.

In his article 'Amphigory and Other Nonsense', van Gelder describes how nonsense and speech associated with madness appear and are theorised in Arabic literature of this period, but admits that a full exploration of the various kinds of nonsense found in this literature, let alone of nonsense as a trans-cultural category, would be too vast an undertaking. There are many ways words can fail to make sense, he explains, from syntactical absurdity to nonsensical narrative sequences, and he provides examples of many of these types of nonsense. It is unclear to me how the poem about Umm Razīn fits into any of the categories that he describes. Perhaps it is described as raving or nonsensical because it does not directly pertain to its context in the story, and also because it describes an absurd action justified with a 'logical' explanation (not unlike ʿAlī ibn Ṣāliḥ's moralising speeches made to justify his standing on one foot waiting for sugar in the middle of the night).

Abū al-Qāsim is also said to rave when he remarks, 'This is odd manners, gentlemen! This is the uncouth language of Baghdad's Bāb al-Ṭāq, and the strange whims of chance!'[35] Though I confess that I do not know what this means, it does not sound especially like nonsense or raving to my ears. This latter example of 'nonsense' is disturbing to me because it makes about as much sense to me as does the rest of the *Ḥikāya*. From our distance in cultural time and space, it is difficult to understand why we do not understand this party-crasher and his comments. For no doubt the confusing quality of his conversation is due in part to the specific references to time and place found in some of the vocabulary, which references are lost on the modern reader. Why, for example, he refers to the Bāb al-Ṭāq neighbourhood of medieval

Baghdad is now less than clear. Some of Abū al-Qāsim's conversation during the chess game in particular can be identified as popular sayings and are locatable in books of Arabic proverbs. One senses that most of his comments in this section are in fact similarly derived from idioms and sayings, and naturally many of these sayings are obscure to modern readers. However, as shown previously, much of the party-crasher's conversation throughout the day also seems to confuse his fellow guests and contemporaries as well.

Even if we do not know why we do not understand Abū al-Qāsim, we can know that by emphasising the ideas of 'nonsensical' or 'raving' speech during the chess game, the narrative aligns game-playing with a particular kind of speech. And although the *Ḥikāya* is generally brim-full of difficult language whose meaning often seems impossible to determine, the chess game is one of the most difficult passages to understand. Not only is it filled with forgotten sayings and obscure (nonsensical?) poetry, the chess moves described are rarely clear. Like the rest of the narrative, this game is presented in a kind of hypothetical grammatical construction, with each move described only as a possible example of a move that could be made.[36] This chess game itself is purely hypothetical, for when Abū al-Qāsim asks to play chess or backgammon, the narrator writes, 'They would bring out, for example, a chess game', suggesting that they could have just as easily brought out a backgammon board.

For each hypothetical chess move, the narrator suggests that Abū al-Qāsim would have a quip or poem to recite. Given that there are more possible chess games than atoms in the observable universe,[37] this would indeed render Abū al-Qāsim a man of microcosmic verbal ability. As for the chess moves themselves, they often seem hastily-described excuses for the production of creative taunts (or in other words, set-ups for a joke). Therefore, just as the portrait of Baghdad provided in the *Ḥikāya* is always ambiguous and unreliable, and the nature of its competition with the city of Isfahan flounders in a disorienting flow of words, the precise chess game played in the *Ḥikāya* is impossible to reconstruct.

To demonstrate the futility of this endeavour, I will here attempt (and fail) to do exactly that. In doing so, I will show that two types of games are being played (as defined by Roger Caillois): *agôn* (competition), the relatively orderly game of chess, and *ilinx* (vertigo), the disorienting and dizzying game of *mujūn*. In the chess game here described, the raving quality of the discourse (*mujūn*) drowns out the orderly nature of the competition, which we find un-reconstructible.

According to H. J. R. Murray, the tradition of white (or red) playing first is a modern one, so we cannot determine the colour of Abū al-Qāsim's pieces based on the fact that he plays first.[38] Abū al-Qāsim begins by advancing his pawns (*bayādhiq*). His opponent likewise advances his pawns. Given that the pawns, unlike in the modern game, could advance only one square at a time on their opening move, Abū al-Qāsim's response to his opponent's seemingly modest and conservative opening sounds especially nonsensical:[39]

> Then his fellow player would advance his pawns, and he would say, 'Hey loser, bite by bite so you don't choke! Just two squares at a time, so you don't end in the black! Camel by camel or you'll break the *maḥāmil*![40] I say enough, but he's sneaking up! Your basket won't split, mister! Don't hurry, my lord, hurrying's for tom-cats. He gets two of my pawns for one pawn, now that's a good deal!'

This last comment seems to suggest that Abū al-Qāsim and his opponent have exchanged pawns, and Abū al-Qāsim in fact goes on to elaborate on this exchange, reciting: 'He traded a beard for my shiny asshole.' In terms of sexual value, a hairless asshole may prove more valuable than a beard, as bearded men were not typically considered attractive (to other men). So it seems that Abū al-Qāsim's opponent is slightly in the lead.

Abū al-Qāsim would then *yastaẓhir bi-firzān band* (perhaps: advance the Queen (*firzān*) protected by a pawn). According to Reinhard Wieber, who cites this very line in his *Das Schachspiel in der arabischen Literatur* as an attestation to the form *firzān-band* (a Persian word), this suggests that the queen's movement is protected by the pawn sitting in front of her.[41] The queen, then known as the *firzān* or adviser, could move only one diagonal space at a time.[42] Abū al-Qāsim, perhaps, advances his *firzān* behind a diagonal line of pawns who had already begun their advance.[43]

Having thus moved his queen in relation to his pawns, Abū al-Qāsim makes a baffling comment: *Iṣʿad bi-liḥāf wa-nzil bi-mirwaḥa.* Cooperson suggests the reading 'Go up with a blanket and come back down with a fan', speculating that, given the pre-modern Arab habit of sleeping on the roof on warm evenings, to go up with a blanket to cool off would be counter-intuitive. Likewise, to come back down, presumably to warm up, while carrying a fan, would also be counter-intuitive. This reading makes this comment another kind of 'discourse that denies itself', providing contradictory, self-cancelling advice. St. Germain, citing a series of Arabic chess terms provided by Wieber, translates this comment as referring directly to the game, though her interpretation of these terms and the resulting translation ('back up a square to go around a piece') I find less than convincing.[44] Cooperson's interpretation seems to me the most likely, given the density of proverbs recited during the chess game, many of them locatable in Freytag's and others' dictionaries of Arabic proverbs.[45] This does not, however, mean that the chess-related double meaning of some of these words is without significance, especially considering that the verbs 'go up' and 'come down' were used specifically to describe the movement of the queen.[46] Ultimately, the meaning is unclear. As with so much of Abū al-Qāsim's abundant and head-spinning speech, we must ask ourselves if even his opponent or the other guests watching the chess game know what he means. This impression is strengthened by the narrator repeatedly describing Abū al-Qāsim's speech during the chess game as 'raving' (*hadhāyān*) or saying that he 'talks nonsense' (*yuhjir*).

The narrator goes on to say that Abū al-Qāsim would limit his opponent's play from the sides.[47] If the opponent broke out of this trap, he would recite an obscene poem. This move is clearly too vague and hypothetical to reconstruct precisely. The narrator continues with a blessedly clear description of the next move: 'His opponent would send his knight into the centre after the advancing of the pawns.'[48] So despite all the talk and braggadocio, we appear still to be in a very early stage of a game, which is unfolding in a usual way with the advance of pawns and a knight (*faras*).

Abū al-Qāsim then suggests that his opponent move 'the king's guard', probably the pawn in front of the king (*shāh*).[49] From this comment we can guess at least that this pawn has not yet moved. Little beyond this is clear. His opponent considers and rejects several moves and is duly mocked for his hesitation. Abū al-Qāsim then takes a pawn on the side of the board,[50] remarking, 'if you can't find a rose, take a cyclamen'. His opponent in turn takes one of his pawns. When the audience asks Abū al-Qāsim why he did not see the threat to his pawn, Abū al-Qāsim tells them to go to hell (in so many words) and takes his opponent's queen or knight. 'A blow of the stone hammer is better than a thousand strikes of the mallet', he says. The next several moves are described in only the most general terms, Abū al-Qāsim's opponent erring and hesitating and receiving due mockery, being forced to take one move he quickly regrets. The opponent moves the knight to the side of the board and is mocked for this by Abū al-Qāsim (it is indeed poor strategy to limit the knight's movement in this way).[51] He then blocks Abū al-Qāsim, who screams (he screams and even neighs several times during the game). But he soon takes some of his opponents' pieces in retaliation. Beyond this, our picture of the state of the board is now wholly unclear. More hesitations and retractions are described than actual moves. At this point we know only that Abū al-Qāsim is dominating both the board and the conversation.

Several more hesitations and false starts follow, until suddenly Abū al-Qāsim's opponent simultaneously threatens his king and his bishop (*fīl*) (perhaps using a 'pin' or 'skewer', as the attack is known in modern chess).[52] It looks bad for our (anti-)hero, who screams again, but in a series of partially nonsensical but clearly threatening poems, indicates that his opponent has in fact fallen into a trap. Indeed his next described move threatens his opponent's rook (*rukh*) and king with his bishop, 'an admirable, elegant move'.[53]

Abū al-Qāsim's opponent struggles for the rest of the game. The narrator tells us that his head is spinning, he curses, his king is in a tight spot, and his pawns are scattered. Or as Abū al-Qāsim puts it, 'He's in shit up to his throat and the dogs are standing guard.'[54] He then moves in for the checkmate using his knight, and upsets the chessboard in raucous triumph, thus ending the game. Or that is my interpretation; in St. Germain's translation, the opponent takes Abū al-Qāsim's knight and overturns the chessboard himself. As reflected in my translation above,[55] the language here is vague, and the winner of the game itself is consequently a matter of uncertainty.

Like so much of the *Ḥikāya*, this chess game is a mixture of tantalizingly specific-sounding details comprising a wholly hypothetical and historically unreconstructible scene, couched in familiar but often incomprehensible conversation. In his work *The Most Human Human*, Brian Christian compares the game of chess to a conversation. In both chess and conversation, there are standard traditional openers and closers from which we rarely deviate. Computers can hold a conversation or play chess within these set parameters as well as humans can. As with life itself, he writes, which always starts with birth and ends with death, it is the variations in the middle that make each conversation or chess game uniquely human.[56] By entering and exiting the party dressed as an ascetic and reciting pious poetry, Abū al-Qāsim evokes the standard pious openings and endings of contemporary Arabic texts. The centre of the *Ḥikāya*, however, casts jarring doubt on even these familiar signposts.

On 27 January 2002, Bobby Fischer, former chess World Champion, declared the game of chess dead.[57] Entire games could now be played 'by the book', a phrase used to describe moves memorised in patterns, themselves repetitions of former games by past masters. Though filled with literary clichés – familiar tropes and poetic quotations – the *Ḥikāya* is nevertheless not a dead game, and not 'by the book'. Partly by virtue of its hypothetical language, which leaves all events described hovering between the realms of truth and fiction, and partly by virtue of Abū al-Qāsim's refusal to adhere even to the outlandish norms of the buffoon (he will not let his audience laugh), the *Ḥikāya*, like the chess game described within it, remains stubbornly anomalous – resembling those games recorded in 'the books', but not quite identifiable within these books. So *mujūn* may be a crazy game, but it may tend to occur within certain temporal and literary parameters in the game of daily life which it parodies and inhabits. *Ḥikāyat Abī al-Qāsim*, baffling, evading, and turning the very game-board on its face, does not play by the rules.[58]

Notes

1. For examples of such usages of the *Ḥikāya*, see Aḥsan, *Social Life*, p. 11, and Dhū al-Nūn Ṭāhā, 'Mujtama'' pp. 14–25.
2. Meisami, 'Arabic *Mujūn* Poetry: The Literary Dimension', p. 9.
3. Szombathy, *Mujūn*, pp. 34–42, 303–9. Meisami reviews some earlier attempts to define *mujūn*, and ultimately calls it 'bacchic and erotic, both taken to extremes … not merely anti-religious but also anti-heroic in outlook'. She later adds that *mujūn* is a 'counter-genre which inverts the conventions of "normative" *ghazal* and *waṣf al-khamr*' (Meisami, 'Arabic *Mujūn* Poetry', pp. 8–9, 14). Van Gelder points out that in distinguishing it from a different style, *hijā'*, or invective, 'the

text alone is not enough to decide its nature: one needs to know the intention of the poet and his relationship with the victim' (Van Gelder, *The Bad and The Ugly*, p. 51). He is distinguishing between invective (*hijā'*) which, in Rowson's words, has 'defamatory intent', and *mujūn*'s 'essential lightheartedness'. Rowson later adds 'It is less the illicitness of the subject than the presence of explicit vocabulary and graphic description that sets off *mujūn*. Rowson, '*Mujūn*', pp. 546–7. Many further definitions and explorations of *mujūn* are undertaken by multiple authors in *The Rude, the Bad and the Bawdy*, edited by Adam Talib, Marlé Hammond, and Arie Schippers (Gibb Memorial Trust, 2014).

4. Al-Azdī, *Sittenbild*, p. v.
5. Al-Zubayr ibn Bakkār, *Akhbār*, p. 69.
6. Leder, 'Prosa-Dichtung', pp. 6–41.
7. Examples of this are found throughout Victor Turner, *The Ritual Process*, which discusses the balance between communitas and hierarchical structures in society.
8. Kurke, *Coins*, pp. 256–60.
9. Kurke, *Coins*, pp. 268–70.
10. Kurke, *Coins*, p. 263.
11. See pp. 122–3.
12. For a description of similar nicknames used for food in Arabic literature, see van Gelder, 'Edible Fathers and Mothers', pp. 105–20.
13. Al-Masʿūdī, *Murūj* I: 161, V: 3477–81.
14. Morley, *My One Contribution*, p. 47, quoting Jane Harrison in *Ancient Art and Ritual*.
15. Morley, *My One Contribution*, p. 78.
16. This poem, by Badr al-Dīn ibn al-Ṣāḥib, is quoted by al-Ghuzūlī in *Maṭāliʿ al-budūr* and translated by Franz Rosenthal in his *Gambling in Islam*, p. 161. It echoes Qur'anic descriptions of the Day of Resurrection, on which day God will revive the bones (*ʿiẓām*) of the dead to be judged (e.g. Q 75:3).
17. Kilito, *Les Séances*, pp. 228–32.
18. Rosenthal, *Gambling in Islam*, p. 89.
19. Many opinions about chess were attributed to early authoritative figures in the history of Islam during later periods in which the game was far more widespread and popular.
20. Al-Ājurrī, *Taḥrīm al-nard wa-l-shaṭranj wa-l-malāhī*, p. 61. The title of this tenth-century work means 'The prohibition of backgammon, chess, and games'. ʿAlī is further discussed in Chapter 5, pp. 151–2.
21. Van Gelder's translation, based on the reading of the word *al-ʿawāl* as 'chess

master' (as in Wieber, *Schachspiel*, p. 318), though a pun with its other meaning of 'long lances' is clearly also intended.

22. Trans. Robson, 'A Chess *Maqāma*' pp. 114–15.
23. Ibn Abī Ḥajala, *Unmūdhaj al-qitāl*, pp. 137, 59–62.
24. See Murray, *History of Chess*, p. 209 for further examples. Also see Kai-khusru, *The Explanation*, pp. 13–14, a Zoroastrian Pahlavi text.
25. G. J. van Gelder suggested the term 'macranthrope' in a private communication to me.
26. This epistle and its relation to the *Ḥikāya* is explored in Chapter 5, pp. 144–6.
27. *Ḥ*, p. 279.
28. Stewart, *Nonsense*, pp. viii, 72–7.
29. Stewart, *Nonsense*, pp. viii, 72–7.
30. Umm Razīn's name means something like 'Mother of the serious one'.
31. Mentioned above in 'Those Camels have Passed' in Chapter 2, p. 92, and below in Chapter 5, p. 137.
32. Al-Nīsābūrī, *Uqalā'*, pp. 39–57.
33. In Chapter 5, p. 137.
34. Especially his seemingly incoherent remarks at the end of the text, discussed in Chapter 5, pp. 157–8, 174.
35. *Ḥ*, pp. 288, 289.
36. For example: 'And if his opponent would take up one of his pawns in his hand, and act as if he were going to move it, he would say, 'If you see the chicken pecking the rooster's ass, you know that she's telling him "fuck, fuck!"' (*Ḥ*, p. 282).
37. This fun fact has widely circulated among chess and math enthusiasts, and is explained in detail in Breslin, 'Number of Possible'.
38. Murray, *History of Chess*, p. 224.
39. Murray, *History of Chess*, p. 226.
40. Litters for camel-back, which presumably could be become entangled and break if the camels walked side-by-side.
41. Wieber, *Schachspiel*, pp. 322–3.
42. Murray, *History of Chess*, p. 225.
43. Khawam translates: *Cette fois, c'est la Reine qu'il envoie en renfort sur le pion le plus menacé*, suggesting that Abū al-Qāsim moves a queen in order to defend a pawn, though this seems an unlikely strategy, even given the diminished value of the queen in medieval Arabic chess (al-Azdī, *24 heures*, p. 245). It is tempting to consider this a reference to a set opening series of moves (see e.g. Wieber, pp. 350, 351, 354, 355, 356, 357).

44. St. Germain, *Anomalous*, p. 328. Wieber, *Schachspiel*, pp. 308, 333.
45. Adam Talib suggested in a private communication to me that this potential proverb could be read as meaning 'cover all of your bases', or bring a blanket even if the weather is warm, and a fan even if it is cool.
46. Murray, *History of Chess*, p. 227.
47. St. Germain interprets this line not as a move but as further commentary of Abū al-Qāsim (*Anomalous*, p. 328). But it seems unlikely that Abū al-Qāsim is speaking here because this comment is *followed* by 'he would say', which throughout the game introduces Abū al-Qāsim's speech, while commentary of the audience is introduced by 'it would be said'.
48. *Ḥ*, p. 281.
49. Or perhaps a piece otherwise protecting the king from attack (see Wieber, *Schachspiel*, p. 300).
50. *Ḥ*, p. 283.
51. A chessboard at that time would probably have been made of soft patterned cloth (Murray, *History of Chess*, p. 220).
52. Thanks to Matthew Leigh for informing me of these terms.
53. *Ḥ*, p. 289.
54. *Ḥ*, p. 291.
55. See p. 56.
56. Christian, *Most Human Human*, pp. 99–131.
57. This radio interview with a station in Reykjavik can be found on Chessbase News.
58. An earlier version of this chapter was published in Selove, '*Mujūn*'.

5

The Cosmic Crasher

> O Nature, and O soul of man! How far beyond all utterance are your linked analogies! Not the smallest atom stirs or lives on matter, but has its cunning duplicate in mind.
>
> Captain Ahab in Melville, *Moby-Dick*, 'The Sphynx'

> Seems to me some sort of Equator cuts yon old man, too, right in his middle. He's always under the Line . . .
>
> Melville, *Moby-Dick*, 'The Deck'

Al-Azdī tells us that Abū al-Qāsim is based on a real person that he once knew, though he does not suggest that this friend was himself named Abū al-Qāsim. We may have reason to believe that the name was designed to achieve a particular effect, for it is also the name of the prophet Muḥammad, whose son Qāsim passed away at a young age (the Prophet's *kunya*, or filionymic, means 'Father of Qāsim'). Abū al-Qāsim's *ism* (given name) as it is found at the beginning of the text, Aḥmad (another name for the Prophet, derived from the same root as the name Muḥammad), strengthens this impression.[1] The use of names associated with Muḥammad to designate a man of Abū al-Qāsim al-Baghdādī's qualities may at first strike us as inappropriate, but closer examination shows that the prophet and Abū al-Qāsim share more than a name alone. Both are at once real human beings, and 'the cosmic individual in whom all [the Islamic Community's] faculties are realized.'[2] This is H. A. R. Gibb's characterisation of the prophet Muḥammad in the view of mystical Islam, but it could just as well describe Abū al-Qāsim, introduced as an all-encompassing intersection of all the people of Baghdad.

In many ways Abū al-Qāsim seems a travesty of the prophet, suggesting the distance between the ideal of the Islamic *umma* (community) as embodied in Muḥammad's epic person, and contemporary Baghdadi discourse, which, as it is represented in the *Ḥikāya*, tends towards decadent worldliness. Indeed in the closing passage of the *Ḥikāya*, the author describes his protagonist as 'tantamount to Satan',[3] a comparison whose aptness is described below in the section 'Abū al-Qāsim as Iblīs'.[4] Although this lowly party-crasher is introduced as being 'scrounged out of a garbage-can' and 'a scrap on the manure pile',[5] by the end of the tale it is clear that he is not without his own claims to cosmic significance, and represents the paradoxical joining of the prophet Muḥammad and Iblīs in the unity of a microcosm.

Abū al-Qāsim as Microcosm

Because of Abū al-Qāsim's bad behaviour, his affinity to the Prophet and his reflection of the higher elements of a microcosm are not easy to discern. He is, however, by no means unique in world literature as an uninvited guest dressed in rags, whose visit represents the unexpected visit of the divine. (One famous example of this topos, the tale of Philemon and Baucis in Ovid's *Metamorphoses*, is the subject of a section in my conclusion.) Furthermore, the image of the party-crasher as a mystic who 'knows the unknown'[6] often appears. Al-Khaṭīb al-Baghdādī's *al-Taṭfīl* describes a party-crasher in Basra, the leader of a kind of party-crashing cult, the members of which are able to divine the secrets of the dishes consumed at a party.[7] In the same text, a party-crasher is satirised as having arcane knowledge of hidden things, as long as they are hidden in pots,[8] a joke found also in the *Ḥikāya.*[9]

On a more earnest note, al-Khaṭīb al-Baghdādī's work subtly presents the party-crashers as the allies of the prophet Muḥammad, who himself helped uninvited people get into parties by asking the host outright to invite them. In one such anecdote, for example, the Prophet says to the host of a dinner-party, 'A man followed us who was not there when you invited us. If you give him permission, he will come too.' The host gives him permission.[10] This stands in contrast to a later anecdote in al-Khaṭīb al-Baghdādī's work in which a party-crasher follows a hadith transmitter to a party, and the transmitter, instead of quietly alerting the host to the situation, attempts to humiliate the crasher in front of his fellow guests by reciting a hadith which states, 'He who enters a gathering to which he was not invited and eats their food enters as a thief and leaves as a looter.' The hadith transmitter is himself humiliated, however, when the party-crasher points out that this hadith is not considered authentic, and recites the more reliably-transmitted hadith which states that the prophet Muḥammad said, 'Food for one is enough for two, and food for two is enough for four, and food for four is enough for eight.'[11] Through such stories as these, it becomes clear that the Prophet is an ally of the party-crasher in al-Khaṭīb al-Baghdādī's collection.

Abū al-Qāsim enters the Isfahani party with the appearance of a holy man, and his subsequent mad and blasphemous ravings do not necessarily set him apart from other holy men of his time. In his catalogue of different types of madmen, al-Nīsābūrī describes those who pretend to be crazy to conceal that they are beloved of God, and emphasises that Muḥammad himself was considered a madman and a sorcerer by some of his contemporaries. As one sage puts it, '*man ʿarafa rabbahu kāna ʿind al-nās majnūnan*' (whoever knows his Lord the people think mad).[12] It was not easy to discern who was a madman/trickster and who was a friend of God; indeed one branch of mystical Islam called Malāmatiyya (loosely, Blameworthies) encouraged its members to 'incur public blame by deliberately transgressing the limits of legal and social acceptability', because the praise of one's fellow man is an earthly pleasure unworthy of the sincere believer.[13] Sufi mystics holding various antinomian beliefs were accused of the reprehensible behaviours described or practiced by Abū al-Qāsim, including hypocrisy, sexual promiscuity, wine-drinking, staring at beardless youths, playing games like chess, and being ecstatically moved by music.[14]

Wandering Sufis often excited the suspicion of their contemporaries, who accused them of sponging;[15] indeed some practitioners of mystical Islam, in rejecting worldly things, refused gainful employment and lived off handouts.[16] Abū al-Qāsim himself perpetuates this stereotype, calling a sponger at the Isfahani party a Sufi who asks but does not give in return.[17] Al-Tawḥīdī is thought by some to be the author of the *Ḥikāya* in part because he, like the protagonist of the *Ḥikāya*, seems at once a wise mystic and a foul-mouthed beggar. Further complicating the mystic-rogue dichotomy is the fact that beggars in contemporary Arabic literature frequently adopt a pious front to garner sympathy, as described by a miser in al-Jāḥiẓ's *Kitāb al-bukhalāʾ* (translated by R. B. Serjeant):

> The *muqaddis* (sanctimonious) is a person who stands by a dead man, begging for a shroud for him. He stands on the road to Mecca by a dead donkey or a dead camel, claiming that it was his, averring that by its death he has been frustrated from performing the pilgrimage. He has learned the dialect of the Khurāsānīs, Yemenis and Ifrīqiyyans and become acquainted with the towns, roads and men of those countries. When he wants, he is an Ifrīqiyyan, when he wants, he is of the Farghānah folk and when he wants he hails from whatever district of the Yemen he likes.[18]

This mock holy man bears a striking resemblance to the *ḥākī* (mimic) described by al-Jāḥiẓ in al-Azdī's introduction. Of this *ḥākī*, to whom the author al-Azdī likens himself recreating the speech of the chameleon Abū al-Qāsim, al-Jāḥiẓ writes:

> Nevertheless, we sometimes see a man who can do impressions, of the people of Yemen for example, recreating their unusual pronunciation in every particular. He could imitate a Moroccan, a Khurāsānī, a Persian, a Sindī, or an African, and yes, even to the point that it seemed more natural to him than to them.[19]

These descriptions emphasise the deceptive language of the seeming holy man. These tricksters will say anything, and convincingly, using the authority of their eloquence to gain the most personal advantage out of a situation. They are like the trickster hero of Badīʿ al-Zamān al-Hamadhānī's *Maqāmāt*, who, repeatedly found preaching pious sentiments for cash that he subsequently spends on wine and women, repeatedly justifies his actions with an eloquent verse on the mutability of time.[20] The trickster heroes of the *maqāmāt* are masters of disguise. In Ibn Nāqiyā's tenth *maqāma*, the trickster al-Yashkurī, who is forever adopting disguises, claims to be a prophet whose divine message is 'Wine today, business tomorrow!' [21] Hamori writes that 'the ritual clown and the man of pious exhortation depend on one another', and the *maqāmāt* 'unite the two roles in a single person'.[22] It is often easiest, however, to understand these characters as tricksters whose pious sentiments are simply a ruse.

The *Ḥikāya* shares many passages in common with al-Hamadhānī's *Maqāmāt*, whose hero Abū al-Fatḥ often closely resembles Abū al-Qāsim.[23]

Partly because of this resemblance, some may see Abū al-Qāsim as a low-life hypocrite who adopts the garb of a holy man simply to get fed. However, as an embodiment of a microcosm, Abū al-Qāsim appears on the contrary to genuinely embody both the high and the low, the holy and profane, in all their muddled commingling in a cosmopolitan society, as shown below.

With the *Ḥikāya*, however, we are witnessing more than a Bakhtinian carnivalesque ambiguity. We may well be tempted to turn to Bakhtin, who consistently describes Rabelais's jolly giants as cosmic men and microcosms, and centres of joyful ambiguity. To Bakhtin, the contradictions of praise and blame, old and new, high and low (which all resonate with the portrayal of Abū al-Qāsim), have a very physical presence. They are located in the body, and symbolise the fertilising powers of death, and the constantly self-renewing mass of humanity. However, although Rabelais's language is, like al-Azdī's, encyclopedic, overflowing, obscene, and often dense, Bakhtin repeatedly argues that this language is representative not only of universal man in an abstract sense, but of the concrete and physical reality of sixteenth-century France. Thus, when he describes Rabelais's work and its characters as a microcosm, he means, in part, that the work contains a miniature geography of France, whose features can be confirmed by historical research. The places around which the giants rampage really existed much as they were described in *Pantagruel*. Though specific place-names in both Baghdad and Isfahan are mentioned in the *Ḥikāya*, they are first mentioned as excuses to play word games with these names, for example, 'I hear names like Sārmana (which means camel shitty), Kalmīrai (meaning goat shitty), Adhār (he comes a-farting in their beards), Kūrasmān (constipation or diarrhoea), Kūrishān (shit in the beard) …'[24] As shown throughout this study, the *Ḥikāya* tends to eschew physical reality in favour of language play. Isfahan, in which the tale unfolds, is described only in improbable hyperbole, and Baghdad, the focus of al-Azdī's portrayal, is often more concealed than revealed by the profusion of Abū al-Qāsim al-Baghdādī's speech.

And yet in his introduction, the author al-Azdī promises us a written mimesis of all Baghdadis, organised in a single form (*ṣūra wāḥida*). By way of explanation, he quotes al-Jāḥiẓ's description of a mime, who produces the platonic ideal of that which he mimics in his performance.[25] Because of his capacity to mimic all things, al-Jāḥiẓ adds, the ancients claimed that man is a microcosm,

> because he forms every form with his hand, and imitates every sound with his mouth, and because he eats plants like beasts of burden, and eats meat like beasts of prey, and eats seeds like birds, and because in him are the shapes of all the kinds of animals.[26]

Thus the author presents to us his microcosm of Baghdad, his introduction a light rehashing of the language of contemporary esoteric thinkers and their ancient forbears, who are discussed throughout this study. According to these thinkers, the human body was analogous both to an entire city (which likewise has a head, limbs, a soul, and a system for the evacuation of excrement), and to the universe itself, the very orbits of the planets reflecting various physical and spiritual systems found in each individual.

At al-Jāḥiẓ's time, the idea of a man-as-microcosm was by no means a new one, as evidenced by his citation of the 'ancients' (*al-awā'il*). The fine line between men of greatness and foolish madmen, also suggesting in its way the complexity of the human spirit, was similarly ancient in its origins. The famous opposition of the Cynic Diogenes and Alexander the Great is one example of a type of doubling that echoes through the Arabic philosophical tradition, which abounds with narratives of the ancient Greek Cynics, depicting conversations between emperors and ragged ascetics.[27] As Barton writes, 'the fool – the *fatuus*, the *morio*, the *sannio* – are mirror images of the prince, as Diogenes the Cynic was the mirror of Alexander.'[28] Throughout her work she describes the interaction and mingling of these two figures, both marginal and central, in the classical traditions. Even Jesus, she writes, was represented by the *stupidus* character in early Christian mimes.[29] Elsewhere he was compared to the Cynic Diogenes.[30] Likewise the 'fool for Christ's sake', partly modelled on the character of Diogenes,[31] is reminiscent of some models of wise-foolishness in the Islamic tradition described by al-Nīsābūrī (see above).[32]

Philosophers from Plato to al-Fārābī and beyond emphasise the importance of a mediator figure, neither here nor there,[33] and it is chiefly with this liminal category of being that we are dealing here. Especially common in mystical Islam and Sufism was the notion of a saint or prophet, such as

Muḥammad, as representative of the entire community, at one with God and man alike, and a mediator between the two. Though every man is a microcosm, these special people's spirits were especially in tune with the universe at large.[34] Ibrāhīm al-Jīlī's (d. 1424) *al-Insān al-kāmil* (the 'Perfect' or 'Total' Human), though written several centuries after the probable date of the *Ḥikāya*'s composition, provides an illuminating exploration of the ancient concept of man-as-microcosm as realised in Islamic mysticism (and as further discussed in the section 'Abū al-Qāsim as Iblīs', below).[35] Like Abū al-Qāsim, a 'gathering of the beautiful and the hideous',[36] this 'total man' represents the meeting of opposites, or as al-Jīlī puts it, 'a bridge between the divine and the world', both 'transcendent' and 'immanent'.[37]

Although the comparison of Abū al-Qāsim, a dirty foul-mouthed drunk, with this mystical total man may seem inappropriate, according to al-Jīlī, 'What is ugly has its due place in the order of existence no less than what is beautiful, and equally belongs to Divine perfection.'[38] Everything in the universe, hideous and beautiful, forms a part of this divine plan.

In al-Jāḥiẓ's description of man-as-microcosm, man's bad qualities are mixed generously with the good. Elaborating on the passage quoted by al-Azdī, in which man is introduced as a microcosm,[39] al-Jāḥiẓ explains in his *Book of Living*:

> Don't you know that man … is called 'microcosm', progeny of the macrocosm, for no other reason that one finds in him all the forms that exist in the macrocosm? … They have found in him the aggressiveness of the camel, the pouncing of the lion, the treacherousness of the wolf, the cunning of the fox, the cowardice of the [nightingale], the hoarding of the ant, the artfulness of the caterpillar, the generosity of the cock, the sociability of the dog, and the homing instinct of the pigeon …[40]

Al-Jāḥiẓ, who derived many of these comparisons from Arabic proverbs,[41] portrays man as a mixture of the positive and negative qualities found in the rest of creation, with the Arabic language serving as a kind of witness in support of a philosophical truth. In a passage bearing a close resemblance to al-Jāḥiẓ's description above, Abū al-Qāsim describes an ugly Isfahani woman as sharing the worst qualities of things that have otherwise positive connotations:

> By God, there's not a praiseworthy beautiful thing in existence of which she does not share some feature: with the moon, she shares a spotted face; with the pearl, its oyster; with a coin, smallness and yellowness; with a cloud, gloom; with a lion, halitosis; with a rose, its thorn; with a donkey, its braying voice; with a flame, smoke and pain; with a camel, its teeth; with a bull, clumsiness of tongue; with a peacock, its legs and its shriek; with a leopard, its personality and unsociability; with water, scum and turbidity; with the tiger, aggressiveness; with wine, the headache; with homes, the toilet and the cistern.[42]

By including not only animals, but heavenly bodies, water, and fire, Abū al-Qāsim's insults actually paint a fuller picture of man-as-microcosm than al-Jāḥiẓ's animal-centred musings cited above. Al-Jāḥiẓ elaborates:

> This does not mean that he turns into a camel because he can find his way in the desert or because of his jealousy, his aggressiveness, his malice, or his endurance in carrying heavy loads. Resembling a wolf in the extent to which he is predisposed to being treacherous and cunning like it, or his sense of smell, its ferocity and its great slyness, does not imply that he is a wolf.[43]

In this passage al-Jāḥiẓ explains how a man, as a microcosm, can contain many conflicting elements. Nor is it necessary to designate one aspect of this man as his true nature. The true nature of man, as al-Jāḥiẓ explains, is to contain elements from the rest of creation, even if those elements seem to contradict. He goes on:

> So they made him a 'microcosm', since in him are all its elements, its constituents, and its natural qualities. Don't you see: in him are the natures of anger and serenity, the concepts of certainty and doubt, firm belief and aporia, astuteness and stupidity, integrity and guile, sincere advice and fraud, loyalty and treachery, dissimulation and sincerity, integrity and hate, seriousness and jesting, miserliness and generosity, thriftiness and extravagance, humility and pride, sociability and shyness, rashness and deliberateness, discernment and haphazard judgment, cowardice and courage, prudence and neglect, wastefulness and parsimony, vulgarity and dignity, hoarding and trusting (in God), contentment and greed, desire and

> asceticism, wrath and serenity, endurance and loss of control, remembrance and forgetfulness, fear and hope, ambition and despair, blamelessness and culpability, doubt and certainty, modesty and shamelessness, discreetness and indiscretion, confession and denial, knowledge and ignorance, injustice and fairness, boldness and flight, rancour and placability, irascibility and placidity, joy and sorrow, pleasure and pain, expectation and wishful thinking, persistence and repentance, stubbornness and capriciousness, inarticulateness and eloquence, speech and dumbness, determination and hesitation, heedlessness and astuteness, forgiveness and vindictiveness, free-will and nature, and a countless, unlimited number of other qualities.[44]

Among the 'other qualities' al-Jāḥiẓ suggests are also contained in man we will certainly find those used to describe Abū al-Qāsim in the opening lines of the *Ḥikāya*:

> Praising and speaking ill, clever and foolish, noble and base, close and distant, calm and violent, a true friend and a hypocrite, a late-night chatter and a gambler, a top and a bottom ... righteous and heretical, an ascetic and a debauchee, an honour and a shame.[45]

As Balda points out, even the physical description of Abū al-Qāsim suggests a contrast between his venerable white beard and his alcoholic red face.[46] After further emphasising that he frequently insults people, the description continues: 'He was a scroll in a little box in a saddlebag in a (lonely) tower, sealed with ambergris, swathed in green silk.'[47] Although this mysterious passage was constructed mainly around the pleasing assonance of the words 'scroll', 'box', 'saddlebag', and 'tower' (*darj fī durj fī khurj fī burj*), and though it is followed immediately with the descriptors 'a bucket of sin and a bag of mange',[48] it is nevertheless significant in its contribution to Abū al-Qāsim's cosmic persona. As Balda explains, the beard-hairs of the prophet Muḥammad were similarly preserved in tubes wrapped in green cloth and kept as relics,[49] much as saints' body-parts are kept in cathedrals as objects of veneration. Like these cast-off scraps of the prophet's person, that nevertheless contain his sacred essence, Abū al-Qāsim is a seeming scrap of humanity containing a universe of symbolic power.

Before al-Jāḥiẓ's list of opposing adjectives, he mentions several cosmic implications of man's microcosmic status:

> They called him the microcosm because they observed that he creates the shapes (*yuṣawwir*) of every thing with his hand and with his mouth imitates every sound. And they said, 'And it is because his limbs are apportioned among the twelve Signs of the Zodiac and the seven planets. Moreover, he contains yellow bile, the product of fire, black bile, the product of earth, blood, the product of air, and phlegm, the product of water. So the four pivotal points (*awtād*) are based upon his four natural elements.'[50]

Here al-Jāḥiẓ briefly lays out some of the philosophical/scientific explanations for man's status as a microcosm, which can be found in much greater detail in mystical philosophical texts such as the *rasā'il* (epistles) of the Ikhwān al-Ṣafā' (the Brothers of Purity).[51] This was the name by which a group of anonymous authors composed their famous fifty-two epistles, probably in the late tenth century. These epistles are philosophical/mystical expositions of various bodies of knowledge, from music, to geography, to magic. Although they are often described as conforming to a Shīʿī Ismāʿīlī world-view, they freely draw from ancient Greek philosophy and other ancient sources of wisdom. Probably composed and refined over a long period of time by multiple authors,[52] these epistles elaborated more fully than ever before on the by-then-already-ancient theory of man-as-microcosm, explaining how each organ and attribute of the human being is found mirrored in landscapes, cities, planets, elements, and other features of the universe. For example, as George Conger describes in his *Theories of Macrocosms and Microcosms in the History of Philosophy*, the Ikhwān liken 'mountains [...] to men's bones [...] rivers to intestines [...] cultural centers to the front of bodies, wilderness to the back.' They drew parallels between 'the wind and breathing [...] winter and old age [and] "stars standing still" to stagnation in men's work.'[53]

The thirty-fourth *risāla*[54] of the Ikhwān al-Ṣafā first uses several explanatory metaphors to show how a unity can be composed of many disparate parts. For example, a city is composed of many different buildings and people, but ruled by one king.[55] Thus, they variously demonstrate their argument:

> The way and the order of the world and the courses of its events, with all the bodies found within it, regardless of differences in their image, varieties

> in their shape, and changes in their superficial features, function in the way of a single human body or a single animal body.[56]

In defending this claim, the *risāla* repeatedly refers to the unity of disparate parts *maʿa ikhtilāf ṣuwarihā* (regardless of differences in their forms). Their language is echoed in al-Azdī's introduction to the character Abū al-Qāsim, who he says represents to him a token of the nature of the people of Baghdad *ʿalā tabāyun ṭabaqātihim* (despite differences in their rank), *wa-kaʾannamā qad naẓamtuhum fī ṣūra wāḥida yaqaʿu taḥtahā nawʿuhum* (and it is as though I organised them into one form containing all their variety).[57] As a microcosm, Abū al-Qāsim contains many various characters within himself. As the Ikhwān say:

> In this city are men and women, old and young, and children ... the good and evil, wise and ignorant, righteous and corrupt, and as these peoples differ in natures, characters, opinions, deeds and customs, likewise there are many souls in the macrocosm.[58]

They then go on to explain that this macrocosm is circular in structure, since the seasons of the year occur in a cycle, and water cycles between the clouds and the sea and rivers.[59] The *Ḥikāya* itself has a cyclical quality, beginning and ending in the morning, with Abū al-Qāsim reciting exactly the same pious phrases to greet both days.[60]

The end of the Ikhwān's epistle especially resembles al-Jāḥiẓ's description of the microcosm because it focuses on animals. As proof that all the seemingly disparate elements in this world are part of the same whole, they describe how some of the higher animals resemble people, while the lower people resemble animals. Elaborating on the idea that there is a system of degrees (*martaba*) in the universe, reflected in the order of planets, they describe a hierarchy in the relation of plant to animal to man. The highest of all plants, the date palm, resembles animals in that it bleeds and dies when decapitated, and has shallow roots that derive nutrients in a worm-like fashion.[61] Likewise the higher animals, like the monkey or ape, resemble man in the shape of his body, causing him to mimic the spirit of man in the actions of his spirit (*ṣārat nafsuhu tuḥākī afʿāl al-nafs al-insāniyya*). Likewise the horse resembles man in his nobility and perseverance. Thus the Ikhwān's description of the higher animals focuses on two animals also important to the *Ḥikāya*, the monkey and the horse.

In the *Ḥikāya*, the monkey appears many times as a grotesque mirror-image of man. First, when insulting the guests, Abū al-Qāsim remarks:

> *These people look like monkeys, but*
> *the monkeys are more charming.*[62]

Next, after insulting the tanbur player, he reproaches the crowd for befriending such a poor musician, saying: 'A monkey looked in a toilet and said, "This mirror suits my face."'[63] Soon after, when spying two friends together, he repeats the insult, saying: 'A monkey looked in a toilet and said, "This mirror suits my face alone!"'[64] One is reminded of al-Jīlī's description of the perfect man, 'who as a microcosmos of a higher order reflects not only the powers of nature but also the divine powers "as in a mirror."'[65] The guests at the Isfahani party, in Abū al-Qāsim's world of rhetoric and insult, display the qualities of man, reflected in an animal form, reflected in a grotesque puddle of filth. Al-Azdī effectively takes the idea of man as a reflection of the sky above and dives with it into the sewer.

The monkey provides the perfect grotesque reflection of man (just as the devil, as fool and trickster, is *simia dei* (the ape of God);[66] as al-Jāḥiẓ writes in his *Kitāb al-ḥayawān*, no animal's face resembles man's more closely than do monkeys' faces. Indeed, he writes, some people can hardly be distinguished from them! Monkeys (some of whom, as described in the Qur'an, may indeed be people monstrously transformed),[67] also laugh, get moved emotionally (*yaṭrab*), eat with their hands, and even mimic (*yaḥkī*). Also, like man, the monkey cannot naturally swim,[68] perhaps suggesting that monkeys share a little of man's distance from the natural world.

When Abū al-Qāsim's monkey-ridden round of insults runs out of steam, he moves on to a lengthy description of horses. This marks the beginning of his comparison of Baghdad to Isfahan; the nobility of a man of Baghdad is reflected in the nobility of his mount, he says, while the man of Isfahan is like 'a billy-goat riding a donkey … a monkey on a nag'.[69] In the Ikhwān's description of the higher animals, they dwell longest on the horse, whose noble nature (*karam*) reflects that of man. Also, the Ikhwān add, the horse can understand the language of man.[70] Language is again made a measure of humanity.[71] In contrast, Abū al-Qāsim calls the other guests beast-like because they cannot understand human language:

> It's like a pack of skittish donkeys, fleeing from a lion – deaf, dumb, and blind, and no brain either.
>
> *What God gave me of intellect*
> *is lost on donkey, sheep, and cow.*
> *They cannot hear me call, nor would*
> *they understand me anyhow.*[72]

One is reminded of the Qur'an's comparison of people who have preserved God's holy books without seeming to understand them, to donkeys carrying heavy tomes on their back (*kamathali al-ḥimāri yaḥmilu asfāran*).[73] In keeping with this tradition, Abū al-Qāsim styles himself as a heaven-sent source of wisdom, and his audience as ignorant animals.

Donkeys in the *Ḥikāya* are the grotesque mirror image of the noble horse, as the donkey dominates Abū al-Qāsim's description of bad horses in Isfahan, as compared to the noble steeds of Baghdad (e.g. 'The donkey is a tripping disgrace, black as ink, like a worn-out waterskin, or a sack of molasses. If the rider stops for company, the donkey shits, and if he lets him be, turns his tail-end to the crowd').[74] But donkeys figure most prominently in al-Azdī's introduction to his *Ḥikāya*, in which he quotes al-Jāḥiẓ's description of a donkey-mimic in Baghdad.[75]

In this passage, al-Jāḥiẓ shows why a mimic is an especially suitable representation of the microcosm, containing the intersection of all creation. As Carlin Barton explains in the case of ancient Roman civilisation, each man may play a part assigned to him in the great theatre of the universe, but the mime plays all parts.[76] In the retinue of *circulatores*, or 'wandering street mimes' (whose homeless milieu seems close to home for Abū al-Qāsim), these entertainers performed in such a way that 'you would believe that many were speaking from one mouth' (as one mime's epitaph boasted).[77] Similarly al-Jāḥiẓ (as quoted by al-Azdī) says of the mime, imitating a blind man: 'It is as though this mimic had gathered the intersection of them all, and contained all of the elements of impressions of the blind in one blind man.'[78]

Al-Jāḥiẓ's *Clarity and Clear Expression*, as quoted by al-Azdī, focuses especially on imitators of donkeys, who seems to produce the platonic ideal of the donkey bray in their mimetic efforts. Bakhtin mentions similar donkey-imitators in his *Rabelais and his World*, citing the 'widespread ass-mimes of antiquity' as proof that 'the ass is one of the most ancient and lasting symbols of the material bodily lower stratum.'[79] They are frequently found in the humorous rhetoric of the Cynics; Antisthenes, for example, compares voting people into leadership to voting to designate donkeys as horses.[80] The Cynic Crates is said to have said, 'We should study philosophy until we see in generals nothing but donkey drivers.'[81] Far and wide, the donkey is found as the comic mirror of man, from Bottom's transformation in *A Midsummer Night's Dream*, to that of Lucius in *The Golden Ass*. In this story, Lucius hopes to transform himself by magic into a bird, but accidentally transforms himself into a donkey instead, and goes on to suffer a series of comic misadventures before returning to his human shape. This transformation is often read as having mysterious Neoplatonic import: just as the soul falls into a lower earthly form, only to regain its exalted status through trial and effort, so Lucius' baser impulses (mainly his greedy, thoughtless curiosity) lead to his initial transformation, while a spiritual awakening, late in the text, returns him to his human state. This reminds us of the hierarchy expressed in the Ikhwān's vision of the microcosm, in which the donkey would represent a lower level of being than that of a man. Similarly, one hadith warns, 'Do none of you fear that if you raise your head before the prayer-leader does, that God will turn it into a donkey-head?'[82]

Prominent among the hardships Lucius faces as a donkey (having been punished for a similar lack of discipline) is his consequent loss of speech. The only sound he can make is the bray of a donkey, among the most ridiculous and off-putting sounds to the human ear. Indeed the Qur'an itself says, 'Be modest in thy bearing and subdue thy voice. Lo! The harshest of all voices is the voice of the ass.'[83] Al-Jāḥiẓ (as quoted by al-Azdī) focuses his description of the mime especially on his mimicking of this despised and comic sound, though he also mentions his ability to mimic forms of human speech, such as foreign accents. Likewise Abū al-Qāsim's speech can be many things, gentlemanly and beastly, but the focus seems to be on the abrasive and laughable. In fact, in the introductory list of adjectives describing his character, two of the first three adjectives describe the harshness and loudness of his voice; he is a *naʿʿār*, and a *zaʿʿāq*,[84] a yeller and screamer, someone who does not heed the Qur'an's advice to 'subdue thy voice'.

Although Lucius manages to transform only into an ass, traces of human beings' magical transformation into all manner of beasts and plants abound around him, so that, when walking down the magic-ridden streets of Thessaly, he cannot look upon a beast or even a stone without imagining that it may have been a man transformed by magic.[85] In Lucius's world too, it is as the 'ancients' say: man contains the potential to be all things. When man is a donkey, his story is a comedy, according to Aristotle's definition, which considers comedy to be a representation of mankind as worse than he is (while tragedy represents him as better and nobler).[86] This definition of comedy seems to fit the *Ḥikāya* well, as Abū al-Qāsim is meant to represent al-Azdī's contemporary Baghdad, but his character is so comically off-putting, the representation could hardly be considered a fair and balanced one.[87]

Comedy, however, does not preclude suffering. Indeed, donkeys' comic effects often result from their suffering. This is certainly true in the case of long-suffering Lucius, beaten and humiliated all across the ancient world. Even Eeyore, who famously remarked: 'Good morning, Pooh Bear. If it is a good morning, which I doubt',[88] is a donkey who amuses with his suffering, not to mention the Freudian/Rabelaisian loss of his tail. Abū al-Qāsim, inveterate brayer, also complains at great length of his suffering. Sometimes these complaints are comic. Just as often though, they seem genuinely sad.

Jung sees the trickster (whom he describes as 'both subhuman and superhuman, a bestial and divine being') as the 'forerunner of the saviour', in part because of his propensity for suffering.[89] Abū al-Qāsim, whether man, god, or beast, is no stranger to suffering: he is seemingly a homeless wanderer, and both his nostalgic yearning for Baghdad and his bitter memories of his hardships there often ring true. Furthermore, his descent into drunken ranting and subsequent rising in the morning as a holy man might remind us of the death and resurrection of a saviour figure. But if so, what is his prophetic message? He seems too bitter and caustic, and too opposed to laughter, to symbolise the sort of joyous renewal Bakhtin sees in the grotesque portrayal of man. Perhaps Abū al-Qāsim's failure to conform to Bakhtin's model arises from his dual status as carnival king and Tappecou (a grouchy old sacristan character). That is to say, Abū al-Qāsim is at once the comically usurping king of the banquet, who dies and comes back to life (or in his case, passes out and rises the next day), and a kind of Tappecou character, an old man 'hostile to laughter',[90] the villain of *Pantagruel*, representing the old guard of power and authority.

Abū al-Qāsim as Iblīs

Above all, Abū al-Qāsim is a man of opposites, and his status as a holy man and his implied resemblance to the prophet Muḥammad is, paradoxically, most easily explained by way of his equal resemblance to Satan. Peter Awn's *Satan's Tragedy and Redemption: Iblīs in Sufi Psychology*, describes the moral ambiguity of the Satan character as represented in various strands of Sufi literature, and first fully articulated in the writings of the revolutionary Persian Sufi mystic, al-Ḥusayn ibn Manṣūr al-Ḥallāj (d. 922).[91] To summarise the narrative briefly, Iblīs (Satan) was a faithful worshipper of God for thousands of years before the creation of man,[92] exceptional for his loyalty and his love for his creator. But when God created Adam, He ordered all the angels, as well as Iblīs, to bow before him. All obeyed except Iblīs. His reason for disobeying is one of the main areas of moral ambiguity in the tale. It seems that he disobeyed because he was too proud to bow before Adam, a mere man, and pride is a great sin. But he also disobeyed because he refused to bow to anyone but God, and thus, in some sense, he remained the most loyal of God's worshippers, and the most perfectly monotheistic in his expression of piety. In more sympathetic portrayals, Iblīs's subsequent exile from the proximity of his creator is portrayed as the agonising separation of the lover from the beloved. His tempting of sinners into Hell is the dutiful enacting of his beloved creator's divine plan, of which he is just the instrument.[93]

The Iblīs of this narrative resembles Abū al-Qāsim in many ways. Both are exiles longing for their home. Both are persuasive speakers, at once learned and wise, and dangerous in their destructive logic.[94] And both are almost buffoons, but not buffoons to be lightly laughed at.[95] In some hadiths, Satan is even characterised as a kind of unwanted dinner guest, who shares your meal when the name of God is not invoked over the food.[96]

Abū al-Qāsim most resembles Satan, however, in his sinful pride. Satan, who proudly refused to bow to Adam, was characterised by saying *anā* ('I am') when he protested *anā khayrun minhu!* (I am better than he is!).[97] In a lengthy, bombastic speech in which Abū al-Qāsim, once again offended by the comment of a fellow guest, tells us who he claims to be, he repeatedly says *anā*, and otherwise identifies himself as Satan:

> Damn you, I am the roiled wave, the complex lock, the fire, the brigand, and the grindstone turning! [...] I am Pharaoh! [...] I am Satan uncircumcised! [...] Do you need anything from Mālik, Treasurer of Hell? Damn you, do you know me? [...] By God's light! An egg from me is worth a thousand; if it were hatched a thousand demons would come out.[98]

His identifying with Pharaoh further confirms his sinful pride; a poem by Rūmī (d. 1273) states, 'The "I" of Manṣūr [al-Ḥallāj] was surely a mercy; | the "I" of Pharaoh a curse. Take note!'[99] Here Rūmī distinguishes between the sinful, prideful utterance of 'I', and the controversial utterance of al-Ḥallāj, who said *anā al-Ḥaqq!* ('I am the Truth!' (i.e. God)). This statement is taken by Rūmī and other Sufis not to mean that al-Ḥallāj was so proud that he thought himself to be divine, but that he was so self-effacing that he lost himself in the divine, and did not distinguish his own ego from the omnipresence of God. And some of Abū al-Qāsim's statements in his 'I am' speech resemble not the 'I' of Pharaoh, but the 'I' of al-Ḥallāj, for example, he also says, 'I am God's power parting the Red Sea! I am the Divine Decree!'[100] He finishes his lengthy rant in a decidedly Ḥallāj-like fashion by saying 'I am I!' (*anā anā!*).

In the Shīʿite traditions, similar speeches were attributed to ʿAlī ibn Abī Ṭālib (d. 661), cousin of the prophet Muḥammad, and considered by those traditions to be his rightful heir and spiritual leader of the Islamic community:

> I am Alexander the Great, mentioned in the old holy books. I am the master of Solomon's seal ... I am the proof of God in the heavens and the earths. I am the cause of trembling. I am the thunderbolt. I am the cry of truth. I am the hour of reckoning to him who denied it. I am the book about which there is no doubt. I am the beautiful names by which God commanded that He be addressed. I am the light from which Moses caught guidance. I am the master of the Trump of Doom. I draw out those entombed. I am the master of the Day of Resurrection ... I look in the face of the Holy Kingship ... I resurrect the dead. I make the rain fall. I light the sun and the

> moon and the stars ... I am living and do not die, for even if I died I did not die ... I am the master of the heavenly bodies. I am the everlasting torment from God. I am the downfall of the ancient tyrants. I am the cleanser of the vicissitudes of time. I am the master of earthquakes and tremors. I am the master of the solar and lunar eclipse. I am the defeater of Pharaoh with this my sword ... I allowed Moses to pass through the sea and drowned the soldiers of Pharaoh. I know the growling of the beasts and the speech of the birds ... I am He who kills twice and gives life twice and He who manifests as I will. I break the tyrants of times gone by and I pluck them out and torture them in the afterlife ... I speak in all tongues. I witness the deeds of all creation, west and east.[101]

Such speeches, which share many images in common with that of Abū al-Qāsim (cited more fully Chapter 1), do not imply that ʿAlī is sinfully prideful, but that he is one with the divine.[102] In reference to this section Mez writes that it indeed derives from a type of religious discourse associated with Shīʿism but was subsequently exaggerated by itinerant beggar-preachers in a manner he describes as 'grotesque'.

Are Abū al-Qāsim's *anā*s the *anā*s of Pharaoh or of al-Ḥallāj and ʿAlī? That is to say, are they sinful expressions of pride, or are they expressions of the mystical effacement of the ego, in which Abū al-Qāsim, as microcosm, melts indistinguishably into the universe and God? And is Abū al-Qāsim like the prophet Muḥammad, as his name suggests, or is he like Satan, as his behaviour suggests?

In his *Kitāb al-ṭawāsīn* (The Book of Ṭā's and Sīn's),[103] al-Ḥallāj extensively explores the concept of opposites, beginning with the assertion that Iblīs is the opposite of Muḥammad. Only through an understanding of the preaching of both Iblīs and Muḥammad, he writes, can gnosis be achieved, for it is in the paradoxical union of these two opposites that the mysterious Oneness of God is found.[104] Abū al-Qāsim is an embodiment of opposites in the person of a microcosm, so his 'I am' speech can and should be read as both pious and satanic. Both readings are spiritually instructive and necessary, for 'one who does not know badness does not know goodness.'[105] As the pivotal mystic thinker al-Ghazālī (d. 1111) reportedly said, 'Whoever does not learn monotheism from Iblīs, he is a dualist (*zindīq*).'[106]

As for the nature of Abū al-Qāsim's religious expression, his 'I am' speech and its resemblance to speeches attributed to ʿAlī is only one example of why the character may be recognised as a Shīʿite. The opening verses he recites upon entry to the party, for example, focus on the martyrdom of the prophet Muḥammad's family in an unmistakably Shīʿite fashion.[107] Later he invokes the day of Ghadīr, the day on which ʿAlī was designated as Muḥammad's successor, according to Shīʿite tradition.[108] It is this resemblance that led Moreh to described the *Ḥikāya* as an attempt to 'mock Shi'ite piety'. [109] But Hamori (in exploring the religious imagery in Abū Nuwās's wine poetry) reminds us that the fact that a type of discourse is borrowed in a humorous context does not necessarily indicate that that type of discourse is being mocked:

> Borrowing may be a cynical leveling of the values in the dominant order ... in turn ... it may be an expression of the kind of synecdochic experience that follows upon perceiving that a mode may be inferior and yet reflect the lineaments of the superior: an experience of correspondence that brings delight and a kind of hope. Borrowing may remind us that a game is afoot, with rules that parody the rules of something serious.[110]

Was the *Ḥikāya*'s borrowing of Shīʿite discourse intended to 'mock Shi'ite piety'? Abū al-Qāsim may be Shīʿite in the expression of his piety, but the other guests' reaction to him do not address the manner of his expression of piety, but the fact that he is (very forcefully) expressing his piety at all, and at a party where 'everybody drinks and fucks'[111] (which is to say, it is not the appropriate time and place for such expressions). And if the *Ḥikāya* is mocking Shīʿite piety, then that would imply that Abū al-Qāsim's character is reprehensible, and thus the *Ḥikāya*'s identification of the character as Shīʿite would impugn Shīʿite piety itself. Abū al-Qāsim is reprehensible and even Satanic, but he is also pious and good, even to the point of being comparable to the prophet Muḥammad. He is both reprehensible and pious at once, just as the city that he represents is at once blameworthy and praiseworthy. To depict these opposites as being One presents an absolutely monotheistic worldview, for to oppose the forces of evil and good as truly irreconcilable opposites would be to support a dualistic worldview in which Good and Evil are two separately powerful forces. Precisely such a world-view was ascribed to the Manicheans, whose very name came to be synonymous with 'heretic' (*zindīq*). Abū al-Qāsim himself, in the beginning of the story, is identified as both a righteous friend of God (*ṣiddīq*), and a heretic (*zindīq*).[112] We cannot identify him as only one or the other.

To say that the *Ḥikāya* mocks Shīʿite piety would imply that it is seeking to exclude Shīʿites as being non-normative. On the contrary, however, the message is one of total inclusion, and not just on the abstract level. The party-crasher is the outsider who seeks inclusion, and Arabic party-crashing stories make the party-crashers the sympathetic protagonists. One *should* include the party-crasher, for to do otherwise would make one a miser, ungenerous with his hospitality (and this is among the most damning qualities possible in a Mediterranean hospitality culture like the medieval Arabic/Islamic world). As detailed above, it can be difficult to distinguish between a trickster and a pious ascetic; indeed it seems that God alone is qualified to distinguish between the two. Therefore, a host faced with an unexpected guest, even a dubious-looking beggar like Abū al-Qāsim, must welcome him to the party or risk playing the miser with a friend of God.

Abū al-Qāsim in his excoriation of the city of Isfahan (which was known for its miserliness, as detailed in 'Baghdad the Party-Crasher'), says the following (and rather suddenly, in the middle of a list of Baghdad's superior incense and glassware):

> I do not see [i.e. in Isfahan] witty, groomed drinking companions, reciting poetry, telling anecdotes, together discussing the fringes of culture.
>
> I only see a party of low, vile sons of boors, base, from stock sunk in laxity of manners and drunken sleep, milling around like cattle in a small pass, disputing over methods and creeds, with Isfahani glass pumpkins in front of them like donkey balls.[113]

Among the bad qualities of Isfahan, Abū al-Qāsim mentions their 'disputing over methods and creeds' (*madhāhib wa-adyān*). Isfahan was known for (in addition to its miserliness), its fervid expression of Sunni piety, and its propensity for sectarian strife.[114] Yāqūt writes that 'ruin had spread' in the city due to the 'discord and fanatical bigotry (*al-fitan wa-l-taʿaṣṣub*) between the Shafiʿīs and Ḥanafīs' (two schools of Islamic law).[115] In al-Hamadhānī's '*Maqāma* of Isfahan', the protagonist is afraid to leave early from an over-long sermon in Isfahan (delivered, as it turns out, by the trickster Abū al-Fatḥ), because he fears 'the savage fanaticism of the people of that place'.[116] One vivid report (though written in the fourteenth century, well after the probable time of the composition of the *Ḥikāya*), describes the city thus:

> The majority of the population is Sunnī, of the Shāfiʿite [sic] sect, and they perform their religious duties very exactly. Most of the time, however, these people do nothing but wrangle and dispute, and here never for a moment is the clash of opposing opinions absent. Hence, when these opposing opinions are rife, all the pleasant conditions of this city of Iṣfahān cannot compensate for the evil results of such strife.[117]

It goes on to cite a hadith of the prophet Muḥammad stating that the Antichrist will emerge from Isfahan.[118]

Abū al-Qāsim also condemns the sectarian tendencies of the city. It therefore seems likely that the *Ḥikāya*, given its message of total inclusion, in which even the cities of Baghdad and Isfahan are held up to be opposites and then later conflated with one another, does not seek to mock the Shīʿites but to include them, along with everything else, in the divine Oneness of God.[119] This concept of divine unity, as expressed in worldly diversity, is beautifully explained in a poem from al-Jīlī's *al-Insān al-kāmil*. Its mention of the Qur'an, a book made of many verses together expressive of divine unity, is especially apposite to our discussion of the *Ḥikāya*, a writerly exploration of unity in diversity, whose relationship to the Qur'an is demonstrated further below.[120]

> *The Unicity is a revelation of the Essence*
> *Which appears as synthesis because of the distinction of my qualities,*
> *All in It is unique and differentiated at the same time,*
> *So admire the multiplicity essentially one!*
> *In It, this one is that one, and that which goes is as that which comes.*
> *It is the Divine Reality* (al-ḥaqīqah) *of the multiplicity*
> *Contained in the Divine Solitude* (al-waḥdah) *without dispersion.*
> *By It all finds itself again in the principle of each thing.*
> *And in this respect the negation* (al-nafy) *is equal to the affirmation*
> (al-ithbāt).
> *The essential 'Discrimination'* (al-furqān) *is Its total form.*

And the multiplicity of Qualities appearing in It is like that of the verses (in the Sacred Book).
Recite it then, and read in thy-self the secret (of His Book);
For it is thee, the 'Evident Model' (al-imām al-mubīn) *and it is in thee that is hidden 'the hidden Book'* (al-kitāb al-maknūn).[121]

The Microcosm is a Man

There is a difference, however, between being a microcosm, infused with the divine, and thinking oneself the centre of the world. The latter trait is human enough, and perhaps the pathos of Abū al-Qāsim's display of arrogance in his 'I am' speech, promptly followed by his drunken loss of consciousness, explains why Gabrieli, for example, in comparing Abū al-Qāsim to his fellow tricksters in the *maqāmāt*, writes, 'Whilst [the *maqāmāt*'s] heroes al-Iskandarī and al-Sarūdjī offer us nothing more than a somewhat monotonous and stereotyped cliché figure of a rogue, al-Azdī's Abu 'l-Ḳāsim is wholly alive',[122] and also that he is 'a character of flesh and bones (even too carnal!)' (*un personaggio in carne ed ossa (anche troppo carnale!)*). Abū al-Qāsim is at once terrifying and pathetic, a seemingly indomitable speaker nevertheless not immune to drunkenness or old age. For although his suffering links him to the saviour figure, leading us into cosmic-scale readings of his persona, these same sufferings simultaneously remind us of his humanity. With Captain Ahab he might protest, 'In the midst of the personified impersonal, a personality stands here.'[123] For even as with a saviour or a prophet, it seems Abū al-Qāsim's status as a microcosm overburdens his frail human frame. As Nicholson writes of Abū al-Qāsim's namesake, 'I need hardly say that Mohammed gave the lie direct to those who would have thrust this sort of greatness upon him: his apotheosis is the triumph of religious feeling over historical fact',[124] the historical facts including his individual, mortal humanity.

Muḥammad and his cousin ʿAlī, though human beings, may well lay claim to greatness. In comparison, does Abū al-Qāsim's exaggerated boasting sound a lot like empty talk? Though the Ikhwān al-Ṣafā' and other philosophers perceived man's status as a microcosm as a physical reality, al-Azdī's microcosm is emphatically a world of words, often markedly out of joint with reality. Al-Azdī does not quote the Ikhwān or any other philosophers in his introduction, but rather al-Jāḥiẓ, a man of letters and an innovator in genres of Arabic prose whose truth value is not emphasised. Al-Jāḥiẓ is famous for pitting two debating partners against one another in literary epistles (e.g. girls vs. boys, or the front vs. the back), much like the *Ḥikāya* matches Baghdad and Isfahan against one another in what ultimately amounts to a rhetorical exercise. In his description of mimics and the microcosm quoted in al-Azdī's introduction, al-Jāḥiẓ focuses on man's ability to imitate speech; indeed his first example of man's microcosmic capacities is his ability to imitate the sounds of all the animals.

Al-Jāḥiẓ's explanation of man's microcosmic status itself creates, like the *Ḥikāya*, a full-to-overflowing outburst of language, filled with lists of contrasting adjectives, and with proofs drawn largely from Arabic proverbs. Readers of al-Jāḥiẓ are always wary of his arguments, which seem as often word-games as sincerely expressed arguments. And by linking mimics, who deceive even donkeys with their talents, to the fundamental microcosmic nature of mankind, al-Jāḥiẓ introduces the idea of deceptive and misguiding language into al-Azdī's microcosm. The deceptive quality of language, however, itself contains the seeds of creative power. The *Ḥikāya* has been read as a development of al-Jāḥiẓ's 'type' literature,[125] or books focusing on types of people, like misers, party-crashers, or wise-fools. Abū al-Qāsim's two-sided, self-contradictory manners and speech at once deny the possibility of such definitive categorisations, and allow the creation of a new, seemingly inimitable type of text.

At the end of the story, Abū al-Qāsim's re-emergence from his drunken stupor is described as follows:

> Then he would end in sleep. On that morning the first thing that would be heard was his yell. And he would say, 'We've woken up and the power is to God! Welcome to a new day and a writer bearing witness! Write: In the name of God the merciful, the compassionate, Abū al-Qāsim ʿAlī ibn Muḥammad al-Tamīmī al-Baghdādī says, I bear witness that there is no god but God alone without partner, and that Muḥammad is His servant and His messenger. Our Lord, we believe in what You sent down . . .' etc.[126] 'In the name of God the merciful and the compassionate, ALIF LAM MIM,

> The revelation of the Scripture whereof there is no doubt is from the Lord of the Worlds.'[127] He would mutter in his speech, then sound out with something from the word of the Almighty: 'Their sides forsake', etc.[128]

Abū al-Qāsim is citing from a sura of the Qur'an, al-Sajda, whose topic is the Qur'an itself, as well as other forms of the word of God (i.e. those sent previously to the Jews and the Christians). The injunction 'write!' reminds us of the angel Gabriel's first command to the prophet Muḥammad, 'recite!' (*iqra'!*)[129] the words about to be dictated to him. If Abū al-Qāsim is a *Doppelgänger* of Muḥammad, the *Ḥikāya* is his Qur'an. A fractal-like world of microcosms, the Qur'an contains totality in each of its fragments.[130] As Norman Brown describes in his *Apocalypse and/or Metamorphosis*, this is why Western readers' reactions to the Qur'an as 'a wearisome confused jumble, crude, incondite [with] endless iterations, long-windedness, entanglement',[131] betray a failure to understand the Qur'an's disruptive literary project. In his chapter 'The Apocalypse of Islam',[132] Brown compares the Qur'an to the similarly dense *Finnegans Wake*, which tends to 'reamalgamerge' the shattered pieces of history, language, and symbolism.[133] Similarly, Montgomery writes of al-Jāḥiẓ's *Book of Living*, 'The totalising impulse, the urge to enfold, rather than simply evoke, all of existence in language, pushes language to the very limit of its expressiveness.'[134] We can hear echoes of the *Ḥikāya*'s reception in reactions to Rabelais's *Pantagruel*, which, like the Qur'an, *The Book of Living*, or the *Ḥikāya*, is an encyclopedic work filled with neologisms and the rubble of the human language. As Bakhtin demonstrates, Rabelais' strange language led critics to complain that his work lacked the unity and coherence of a classical text, characterising it as 'extravagant and unintelligible'.[135] These critics, writes Bakhtin, are able to enjoy Rabelais's work only piecemeal, thus missing its unique, universal, and, to Bakhtin, world-creating qualities. The *Ḥikāya*, similarly criticised for its extravagance, excessiveness, and unreadability, similarly contains a world.

Notes

1. At the end of the story (by mistake?) he is called ʿAlī ibn Muḥammad instead of Aḥmad ibn ʿAlī, *H*, 390. The combination Abū al-Qāsim Muḥammad as a name was considered forbidden by many. See Al-Azdī, *Sittenbild*, p. xxv, n. 1. Also see Wensinck, 'Kunya', pp. 395–6.

2. Gibb, 'Foreword', in *Islamic Cosmological Doctrines*, p. xvi.
3. *Ḥ*, p. 391.
4. See p. 150.
5. *Ḥ*, p. 49.
6. Al-Khaṭīb al-Baghdādī, *al-Taṭfīl*, p. 146.
7. Al-Khaṭīb al-Baghdādī, *al-Taṭfīl*, p. 146.
8. Al-Khaṭīb al-Baghdādī, *al-Taṭfīl*, p. 76.
9. *Ḥ*, p. 64.
10. Al-Khaṭīb al-Baghdādī, *al-Taṭfīl*, pp. 54–9.
11. Al-Khaṭīb al-Baghdādī, *al-Taṭfīl* (2006), pp. 64–5.
12. Al-Nīsābūrī, *ʿUqalāʾ*, p. 30, ascribed to 'Abū al-Qāsim al-Ḥakīm' (probably meaning Isḥāq b. Muḥammad al-Ḥakīm al-Samarqandī, a ninth–tenth century religious scholar). Melville's Ishmael similarly remarks of a shipmate lost at sea: 'He saw God's foot upon the treadle of the loom, and spoke it; and therefore his shipmates called him mad' (*Moby-Dick*, p. 440).
13. Karamustafa, *God's Unruly Friends*, p. 36.
14. See Karamustafa's *Sufism*, 'Antinomians and nonconformists', pp. 155–66.
15. For example, van Gelder translates al-Thaʿālibī's comment that 'the eating of Sufis is proverbial: one may say, "eating more than the Sufis" ... This is because it is their profession to eat a lot. They take big mouthfuls ...', adding, 'Such statements derive from a combination of anti-mystical prejudices and the existence of "false", or at least hedonistic, Sufis' (*Dishes*, p. 100. Also see ʿAbd al-Mawlā's *al-ʿAyyārūn wa-l-shuṭṭār*, in which he identifies the *ʿayyārūn* as Sufis (p. 32)).
16. E.g. the Karrāmiyya. See Karamustafa, *God's Unruly Friends*, p. 30.
17. *Ḥ*, p. 64.
18. Al-Jāḥiẓ, *Book of Misers*, p. 44.
19. *Ḥ*, p. 43, also cited in a full translation of al-Azdī's introduction in Chapter 1, pp. 32–3.
20. In the '*Maqāma* of Wine', he is found preaching a sermon during the day, and drinking at a tavern at night. In the '*Maqāma* of Isfahan', he preaches to trick people out of their money. However, in both the '*Maqāma* of Ahwaz' and the '*Maqāma* of Exhortation', he delivers sermons without any expectation of reward. These *maqāmāt* can be read in English translation by Pendergrast (see Bibliography).
21. Lines attributed to the pre-Islamic poet Imru' al-Qays, translated by van

Gelder, 'Fools and Rogues', p. 50. Ibn Nāqiyya lived in Baghdad in the eleventh century.

22. Hamori, *On the Art,* p. 55. This confusion of the holy man and the madman/vagabond is by no means limited to Islamic or Arabic cultures. In medieval picaresque narrative, the saints are similarly transformed into picaroons 'through a comic *imitation* of holy models, in such a way as to confuse the boundaries of the holy and the profane' (Giles, *Laughter,* p. 75).
23. St. Germain describes these parallel passages in *Anomalous,* pp. 15–18, especially a passage in which Abū al-Qāsim insults a fellow guest, and a poem in which he describes all the nice things that he would like to eat (translated in Chapter 1, p. 48).
24. *Ḥ*, p. 93. These are false etymologies of the Persian place-names.
25. This is also discussed in Chapter 2, pp. 71–2.
26. *Ḥ*, 43–4, quoting al-Jāḥiẓ, *Bayān* I: 69.
27. Gutas, 'Sayings by Diogenes'. As Branham writes, 'Cynic rejection of shame flies in the face of the most basic Christian doctrine', while at the same time, their asceticism garners Christian admiration (*Cynics,* p. 19). Also see Kinney, 'Heirs of the Dog', p. 306. It seems the Cynics were always 'bivalent' characters with a similarly dual reception, in ancient Roman culture and even before (see Krueger, 'Bawdy', p. 225). Their ambiguous nature is intrinsic to their rhetorical position. In exploring questions of Cynic behaviour (does their shocking deportment stem from lack of self-control or from saintly humility that shies from all worldly praise?), and the suspicious reception of even Franciscan monks, whose begging and blaming behaviour resembled that of the Cynics (Kinney, 'Bawdy', pp. 307–8). Kinney's article, which also links the Cynics with professional parasite-jesters and the *scurra* (pp. 299–300), might remind scholars of medieval Islam of similarly suspicious receptions of the Sufis, popular holy men, and other dwellers on the margins of society, some of whom seem indeed to have deliberately exploited this ambiguity. In their rejection of affiliation with any one *polis,* and their simultaneous identification with animals, men, and the gods (whose speech and behaviour were similarly unfettered), the Cynics also possessed a certain universal or cosmic quality (see Moles, 'Cynic Cosmopolitanism', pp. 111–13, as well as the introduction to this highly informative volume in which Diogenes is characterised as a 'citizen of the universe', Branham, *Cynics,* p. 25).
28. Barton, *Sorrows,* p. 140. Plato likewise referred to Diogenes as 'Socrates gone mad' (Branham, *Cynics,* p. 9).

29. Barton, *Sorrows*, p. 168.
30. Krueger, 'Bawdy', p. 229.
31. Krueger, 'Bawdy', p. 231.
32. See pp. 125–6, 137.
33. See Morewedge, 'Neoplatonic Structure', pp. 68–70.
34. See Nasr, *Islamic Cosmological Doctrines*, p. 260.
35. See p. 152.
36. *Ḥ*, p. 391.
37. Reynold Nicholson's translation. *Islamic Mysticism*, pp. 78, 87.
38. *Islamic Mysticism*, p. 85.
39. Translated in Chapter 1 (pp. 32–3) and discussed above in this chapter (p. 138).
40. Van Gelder's unpublished translation of al-Jāḥiẓ, *Kitāb al-ḥayawān* I: 212–15. Jeannie Miller provides an astute reading of al-Jāḥiẓ's discussion of man-as-microcosm and his (consequent) relation and resemblance to various animals in her dissertation, More Than the Sum of Its Parts. See especially pp. 211–15.
41. See, e.g., al-Maydānī's *Majmaʿ al-amthāl* 4: 103 (on the homing instinct of the pigeon).
42. *Ḥ*, p. 207.
43. Van Gelder's unpublished translation.
44. Van Gelder's unpublished translation.
45. *Ḥ*, p. 48. For the complete opening description, including Abū al-Qāsim's initial appearance as a holy man, see Chapter 1, pp. 34–6.
46. Balda, 'Marginalité', p. 374.
47. St. Germain's translation in *Anomalous*, p. 157.
48. *Ḥ*, p. 49.
49. Balda, 'Marginalité', pp. 375–6.
50. Montgomery, *al-Jāḥiẓ: In Praise of Books*, p. 410.
51. Introduced in Chapter 4, p. 123.
52. See Marquet, 'I<u>kh</u>wān al-Ṣafā''.
53. Conger, *Theories of Macrocosms*, p. 49.
54. 'About the meaning of the wise men's saying that the world is a giant human', introduced in Chapter 4, p. 123.
55. Ikhwān, *Rasā'il*, p. 214. Melville similarly describes the unity of the ship under the will of its captain: 'For as the one ship that held them all; though it was put together of all contrasting things – oak, and maple, and pine wood; iron, and

pitch, and hemp – yet all these ran into each other in the one concrete hull, which shot on its way, both balanced and directed by the long central keel ...' (*Moby-Dick*, p. 585).

56. Ikhwān, *Rasā'il*, p. 212. Indeed these esoteric speculations seem to have been borne out in modern scientific inquiries. As mathematician Stephen Strogatz described, human cities follow patterns of organic growth displayed by individual organisms. 'These numerical coincidences seem to be telling us something profound. It appears that Aristotle's metaphor of a city as a living thing is more than merely poetic. There may be deep laws of collective organisation at work here, the same laws for aggregates of people and cells' ('Math and the City').
57. *Ḥ*, p. 43.
58. Ikhwān, *Rasā'il*, p. 217.
59. Ikhwān, *Rasā'il*, p. 220.
60. At the beginning and end of the text, Abū al-Qāsim is described as the embodiment of opposites, and at the beginning and end of the text, he insults each guest individually. At the beginning and end of the text, he is described as wearing a pious hood (the *ṭaylasān*), and reciting pious poetry. Indeed throughout the text can be discerned a symmetrical circular pattern reminiscent of James Monroe's theory of cyclical composition in poetry (see Farrin, *Abundance*), which describes a tendency to return to themes introduced early in a poem again at the end, creating the impression that the poem cycles back to the beginning.
61. Ikhwān, *Rasā'il*, p. 226.
62. *Ḥ*, p. 61.
63. *Ḥ*, p. 68.
64. *Ḥ*, p. 81.
65. Nicholson, *Islamic Mysticism*, p. 82
66. Jung, 'Trickster-Figure', p. 255. As shown in the section 'Abū al-Qāsim as Iblīs', below (p. 152), in the mystical Islamic tradition, he is considered the opposite of the prophet Muḥammad.
67. Q 7:166, 2:65, 7:60.
68. Al-Jāḥiẓ, *al-Ḥayawān* IV: 98.
69. *Ḥ*, p. 126. Samples of these passages are found in Chapter 1, pp. 40–2.
70. Ikhwān, *Rasā'il*, p. 225.
71. See Miller, *More than the Sum of its Parts*, p. 118 ff., for a discussion of humans' and animals' ability to speak according to al-Jāḥiẓ.
72. *Ḥ*, pp. 82–3. Later he again compares a fellow guest to an inarticulate animal,

insulting the Isfahanis for their lack of eloquence: 'He looks like a monkey, or he's from the monkey mould | It's like he's chewing poop, he's such a mumbler', p. 223.

73. Q 62:5.
74. *Ḥ*, pp. 131–2.
75. *Ḥ*, p. 43. Cited in Chapter 1 (p. 33), and discussed in Chapter 2 (pp. 71–2) and in this section above (p. 138).
76. Barton, *Sorrows*, p. 137.
77. Barton, *Sorrows*, p. 138.
78. *Ḥ*, p. 43.
79. Bakhtin, *Rabelais*, p. 78. In fact, Abū al-Qāsim shares with the giants a fondness for the insulting term 'donkey-cock' (*viédaze* in French, *ayr ḥimār* in Arabic (Bakhtin, *Rabelais*, pp. 170, 351, and *Ḥ*, 63)).
80. *Dictionary of World Biography*, p. 89.
81. Branham, *Cynics*, p. 10. Also see Jung, 'Trickster-Figure', p. 259, where Jesus is likened to a donkey. Stubb, in an ill-advised attempt to lighten Captain Ahab's mood, even compares the great Moby-Dick to a donkey, saying of the captain's destroyed whaleboat, 'The thistle the ass refused; it pricked his mouth too keenly, Sir; ha! ha!' (Melville, *Moby-Dick*, p. 582).
82. Al-Jābī, *Ḥamīr*, p. 22. This hadith is found in Bukhārī's collection, vol. 1, no. 660.
83. Q 31:19, trans. Pickthall.
84. *Ḥ*, p. 46. Early in his *Book of Living*, al-Jāḥiẓ considers the sounds that animals make, avowing that he himself can understand some of the expressions of donkeys, dogs, cats, and camels, when, for example, they call out for their young, or spy food. He classifies man alone as *faṣīḥ* (eloquent), because he considers these animal's speech acts to arise without rational intention, as opposed to the speech of man (al-Jāḥiẓ, *al-Ḥayawān* I: 31–3). However see Miller, *More than the Sum of its Parts*, p. 122 f. for a discussion of this idea in which she shows that al-Jāḥiẓ considers animal speech different from human speech only in its degree of complexity, in that animals have less complex concepts that they need to convey, but are still able to express their needs in vocalisations.
85. Apuleius, *Metamorphoses*, 2.1.
86. Aristotle, *Poetics*, 1:II.
87. Mubārak, *Nathr*, p. 419.
88. Milne, *Winnie-the-Pooh*, pp. 72, 84.
89. Jung, 'Trickster-Figure', p. 263. This essay could prove illuminating to the

character of Abū al-Qāsim in many of its observations. It essentially reads the trickster figure as being in between the states of animal and human, thus representing the continuing evolution of the human race, which remains in conflict between two mental states, represented by the past and the future. Malcolm C. Lyons uses Jung's reading in his study of the Arabic trickster in his *The Man of Wiles in Popular Arabic Literature.* He compares the Arabic 'man of wiles' to the Winnebago Trickster (as described by Paul Radin), saying that they 'are on a different level and of a different type from the Arabic quoted here, but, for all that, the psychological background must be investigated', p. 242.

90. Bakhtin, *Rabelais*, p. 267.
91. Awn, *Satan's Tragedy*, p. 123.
92. See Bodman, 'Stalking Iblīs', pp. 258–66, for an account of various incarnations of this narrative.
93. Awn, *Satan's Tragedy*, pp. 85–6, (citing Rūmī's *Mathnawī* for the latter two ideas).
94. Awn, *Satan's Tragedy*, pp. 24, 79–80, 88.
95. Awn, *Satan's Tragedy*, p. 116.
96. Awn, *Satan's Tragedy*, p. 52, citing Muslim, *al-Jāmiʿ al-ṣaḥīḥ* VI: 107–8.
97. Q 7:12, 38:76; Awn, *Satan's Tragedy*, p. 92.
98. *Ḥ*, p. 375. For a full translation, see Chapter 1.
99. Awn, *Satan's Tragedy*, p. 93, citing *Mathnawī* Book 1, 1.3214–16.
100. *Ḥ*, p. 375.
101. This translation is based on manuscript no. 1011 in the Caro Minasian collection of Persian and Arabic manuscripts at UCLA. The same speech can be found in Ramaḍān, *Bughyat al-Ṭālib*, pp. 408–10, with minor variations.
102. Al-Azdī, *Sittenbild*, p. xviii.
103. These are the names of two mysterious letters (in this case, *ṭā* and *sīn*) found at the beginning of certain chapters of the Qur'an; the letters *ṭā* and *sīn* are found at the beginning of Sura 27, The Ant. An example of other mysterious letters (*alif, lām, mīm*) can be found in the Qur'an-inspired conclusion of the *Ḥikāya*, discussed on page 157.
104. Ḥallāj, *Ṭawāsīn*, p. 49, Awn, *Satan's Tragedy*, pp. 123–4. The fact that opposites were capable of existing together in one form was even taken as proof of God's existence (Montgomery, *al-Jāḥiẓ: In Praise of Books*, p. 281, citing al-Khayyāṭ's critique of Ibn al-Rāwandī's critique of al-Naẓẓām's theory of the composition of contraries).
105. Ḥallāj, *Ṭawāsīn*, p. 49.

106. Trans. Awn, *Satan's Tragedy*, p. 133. Awn, who located this passage in a photocopy of a manuscript in the Widener Library in Harvard (titled *Kitāb al-quṣṣās wa-l-mudhakkirīn*), tells us that it is also quoted by Massignon, *Recueil de textes*, p. 96.
107. *Ḥ*, pp. 54–5.
108. *Ḥ*, p. 384.
109. Moreh, *Live Theatre*, p. 96.
110. Hamori, *On the Art*, p. 61.
111. *Ḥ*, p. 55. Karamustafa identifies *ṣiddīq* as one of the terms for a friend of God in *Sufism*, p. 2.
112. *Ḥ*, p. 48.
113. *Ḥ*, p. 176.
114. In a private communication to me, Devin Stewart confirmed that he has encountered ample evidence of violent anti-Shīʿite sentiment in the city of Isfahan around the probable time of the composition of the *Ḥikāya.* I anticipate a fuller and more official report on this issue in his upcoming study of the *Maqāmāt.* Lambton, in her *Encyclopaedia of Islam* entry on the city, further confirms the tumultuous, sectarian atmosphere of the city, though mainly with references to authors who lived a century of more after the probable time of composition of the *Ḥikāya*, e.g. Kamāl al-Dīn al-Iṣfahānī, a twelfth–thirteenth century poet, and the fourteenth-century travel writer Ibn Baṭṭūṭa (*The Travels of Ibn Baṭṭuṭa a.d. 1325–1354*, Hakluyt, ed. H. A. R. Gibb, pp. ii, 294–5).
115. Yāqūt, *Buldān*, vol. 1, p. 209.
116. Al-Hamadhānī, *Maqāmāt*, trans. Prendergast, p. 57.
117. Mustawfi, *The Geographical Part of the Nuzhat al-Qulūb*, p. 56.
118. Attributed to Ibn ʿAbbās. Mustawfi, *The Geographical Part of the Nuzhat al-Qulūb*, p. 56.
119. Inspired by Max Weber's 'Religious Rejections of the World and Their Directions', Karamustafa understands Sufism before and during the probable time of the composition of the *Ḥikāya* to be a synthesis of 'two of [Islamic religiosity's] most powerful subcurrents: asceticism and anarchist individualism' (*Sufism*, p. 25). This synthesis 'bridged the abyss' between private, often world-rejecting inner piety, and public, legalistic external expressions of religiosity, as well as the divide between 'God/all other than God'. It did so by means of 'a powerful reinterpretation of the doctrine of unity (*tawḥīd*) … Whatever God created, in particular this world, had to be accepted … In some

sense, this world too, like the other world, was infused with the Divine' (*God's Unruly Friends*, p. 29).

120. See p. 158.
121. Mentioned above, and here translated by Titus Burkhardt: Al-Jīlī, *Universal Man*, p. 23; *al-Insān*, p. 83.
122. Gabrieli, 'Abū 'l-Muṭahhar al-Azdī'. He goes on to compare him to the characters in Petronius' *Satyricon* or to Spanish *pícaros*.
123. Melville, *Moby-Dick*, p. 533.
124. Nicholson, *Islamic Mysticism*, p. 88.
125. Pellat, 'Ḥikāya', p. 368.
126. Q 3:35, 'Our Lord! we believe in what thou hast revealed, and we follow the Messenger; then write us down among those who bear witness' (trans. Yusuf Ali).
127. Q 32:2, trans. Pickthall.
128. *Ḥ*, p. 390. Q 32:15: 'Their sides forsake their beds, to invoke their Lord in fear and hope, and they spend … out of what We have bestowed on them' (trans. Mohsin Khan).
129. Q 96:1.
130. Montgomery, *al-Jāḥiẓ: In Praise of Books*, p. 268.
131. These are Thomas Carlyle's words, as quoted in Brown, *Apocalypse*, p. 89.
132. Brown, *Apocalypse*, pp. 69–94.
133. Brown, *Apocalypse*, p. 88.
134. Montgomery, *al-Jāḥiẓ: In Praise of Books*, p. 58.
135. In the words of Voltaire; in Bakhtin, *Rabelais*, p. 116.

Conclusion

> Thus we see now that the spine of even the hugest of living things tapers off at last into simple child's play.
>
> Melville, *Moby-Dick*, 'Measurement of the Whale's Skeleton'

Bojangles Won't Dance

In the pages (or hours) before Abū al-Qāsim's superhuman vaunt, he appears to me the most human, or the most 'of meat and bones'.[1] In these pages the narrator informs us that Abū al-Qāsim grows increasingly intoxicated and physically exhausted, and in these pages he interacts with the other guests more than anywhere else in the narrative. This is to say, the other guests do here occasionally get a word in edgewise. On page 334 of al-Shāljī's 391-paged edition of the *Ḥikāya* (*Ḥ*), our (anti-)hero, already well into his cups, and having thoroughly exhausted the game of arbitrary praise and blame, recites a poem in a voice choking with emotion (*ṣawt shajin*) and then wistfully declares: 'Tomorrow, by God, we'll resume this party and its pleasures!'[2] The best way to greet a new day, he explains, is with more drinking and revelry. This suggestion transgresses normative partying behaviour,[3] and sets a theme for the denouement of the *Ḥikāya*, in which Abū al-Qāsim repeatedly strives to extend the pleasures of the party beyond the human capacity for enjoyment, with limited and ambiguous success.

Continuing to drink, Abū al-Qāsim praises a beautiful female singer, and then fiercely insults her guardian, in many ways a typical praise/blame sequence for the protagonist, though earthily connected in this case to the bodies of people in the room and his physical responses to them (his lust for the singer is obvious, and she rebukes him for it later in the narrative). His expression of hatred for her guardian is even more fulsome in its invective than usual, so when Abū al-Qāsim ends with a quick couplet wondering if, perhaps, he has left too much unsaid,[4] a fellow guest understandably bursts into laughter.

In the song Mr Bojangles – whose titular character was, like Abū al-Qāsim, said to be based on a real man[5] – a tragic figure, an avowed alcoholic prisoner perpetually in mourning for his dog, is repeatedly asked to dance by his fellow inmates. And what if he chose to refuse? Abū al-Qāsim, though widely identified by scholars as a popular performer singing for his supper,[6] emphatically refuses to be laughed at. As at the beginning of the text, when Abū al-Qāsim meets a guest's laughter with an outburst of almost unparalleled fury, the laughter at the end of the party again sparks a verbal rampage. Abū al-Qāsim suggests that laughter is inappropriate in the face of so much pain and so formidable an opponent: 'The laughter of a serpent in a bag of lime! A wolf laughing as the hounds close in! A sheep's-head laughing in the butcher shop! ... The laugh of a whore when the midwife rebukes her!'[7]

The guests are baffled by his response, which marks a critical turning point in the narrative. Everyone now decides that they must get rid of Abū al-Qāsim, the non-stop joker who will not be laughed at, and they plot to encourage his growing intoxication to the point that he passes out. Those guests not currently involved in a quarrel with their belligerent fellow diner approach him with goblets of wine. 'Slow down, creatures of God!' he tells them. 'Camel by camel, don't break the *maḥāmil* (camel-litters)!' This is the second time Abū al-Qāsim has used this rhymed expression. The first was during the chess game, after his opponent moved his first piece, a pawn (as described in Chapter 4). In that case his injunction to slow down is comically hyperbolic – his opponent had made only one move, and a seemingly ordinary and prudent one. Nor, when Abū al-Qāsim uses the warning a second time, does it seem at all appropriate, for he cannot be called a man of moderation or timeliness. When he calls on his fellow guests to slow down and practice moderation, he is already struggling to prolong the party beyond its natural lifespan, and encouraging fellow revellers to drink and party even in the morning hours, a suggestion that flies in the face of timeliness.

Bakhtin calls time itself the author of carnivalesque narratives, in which the protagonist is first exalted and then beaten down.[8] Abū al-Qāsim, as the day and the party run out of steam, struggles against this inevitable decline. When he lustily assaults the beautiful slave girl, she rebukes him, saying, 'Don't you know that you're an old man?' Abū al-Qāsim responds with more lusty poetry. Later, as the guests continue striving to subdue him, he says 'The old men become like children!'[9] Soon afterwards they admonish him again, saying, 'How long will this stupidity continue, old man? Have you no shame?'[10] He dances and sings until he falls on the ground in exhaustion. The party will soon be over.

Perhaps we can discern elements of the joyful renewal described by Bakhtin in Abū al-Qāsim's behaviour. Or perhaps he represents Baghdad's unwillingness to forget its former glory long after it had largely fallen into ruin.[11] Hamori reads drunkenness itself in Arabic literature as a struggle to transcend or forget the inevitable passage of time.[12] And although Abū al-Qāsim is eventually conquered by his exhaustion and falls asleep, he wakes up repeating the very same pious words that began the book: 'He would recite the verses in the same order as at the beginning of this epistle, as well as the moralising described there. Then he would don his pious robe just as he first appeared.'[13] He passes out of the lives of the other guests as he leaves their party, but seems, despite his age, changeless and unconquered by time.

Nevertheless, it is during these final pages of the *Ḥikāya* that the protagonist seems most poignantly human, an old and weakening man.[14] We are left wondering if he is human or superhuman, divine, or profane. The author himself certainly does not suggest that any conflicts or questions presented by his protagonist are meant to be resolvable. He ends his *Ḥikāya* saying:

> This is an imitation of Abū al-Qāsim al-Baghdādī al-Tamīmī and his attitudes, which clearly show you that he was the disgrace of his day and the equal of Satan: a gathering of the good and the ugly, transgressing the extreme and the limit, perfecting the conjuncture of seriousness and jest, composed of complete sincerity and complete hypocrisy, created from these things like the people of Iraq. Praise be to God who is One, and His blessing on our lord Muḥammad, His prophet, and on his family, and peace.[15]

His particular mention of the Oneness of God suggests that all of these contradictions can and do coexist within a unified whole. Nevertheless, Abū al-Qāsim's aged humanity does seem actively at odds with his superhuman nature, and unable to keep up with its demands. So is Abū al-Qāsim an archetype or an old man? In his work *Trickster Makes this World*, Lewis Hyde identifies the trickster as a figure of pantheism, whether he be the coyote of the Nez Percé or China's monkey king. Hermes, one of trickster's better-known manifestations, Hyde identifies as a kind of party-crasher of the gods. In the *Homeric Hymn to Hermes*, the newborn baby god steals Apollo's cattle and sacrifices them to all of the deities of Olympus, of whom he will soon be a member. It is a theft that 'confuses the definition of theft'.[16] In this story, Hermes, like Abū al-Qāsim, does not behave appropriately to his age, but seems a 'grey-haired baby',[17] a liminal embodiment of conflicting opposites (Hermes acts far too old for a baby, and Abū al-Qāsim, far too toddlerish for a shaykh). Hyde reads the baby Hermes' theft as one typical of the Trickster, who does not eat the food that tempts him but rather turns it into a symbol. At the end of this story, Hermes earns the forgiveness of Apollo by giving him the lyre he has just invented.

This myth is distant from the *Ḥikāya* in time, purpose, and form, but as an archetypal embodiment of the trickster, Hermes shares some features with Abū al-Qāsim and some of his fellow party-crashers. Hermes' transformation of the stolen meat into a burnt sacrifice parallels the party-crasher's exchange of food for words, at once questioning the true ownership and significance of the food and drawing the interpretive eye to the symbolism of the trickster's act. His invention of the lyre is echoed in the party-crashers' winning artistic creativity: their unexpected visit represents an opportunity for creation and change, and they earn their invitation to the banquet (or in Hermes' case, Olympus) with their wit and eloquence, even as they ensure their own marginalisation with their outlandish behaviour. Homer's hymn even likens Hermes' first song on his newly-invented lyre to the playful invective of a reveller.[18]

Hermes' mischievous acts of invention and thievery bring old assumptions and truths into question, but instead of offering a newly codified 'truth', they encourage interpretation and further change. 'The thief's last theft', Hyde writes, 'is to steal himself away ... he is not the declarative speaker of traditional prophecy.'[19] By removing himself as he does at the end of the party, Abū al-Qāsim creates more questions than answers.

But 'there are no human tricksters' Hyde writes; 'Human beings participate in this mythology, but they simultaneously participate in others, and in history.'[20] Abū al-Qāsim, in his irascible humanity, does not comfortably inhabit his archetype. It is clear from al-Azdī's introduction to his work that a conflict will be staged in the *Ḥikāya* between reality, language, and representation. The protagonist and his fellow guests seem to suffer from this very conflict, as does the reader of the *Ḥikāya* (judging by the complaints of modern-day critics). The author promises to tell us something about Baghdad. The party guests themselves expect Abū al-Qāsim to tell them something about Baghdad. But their informant replies with lists of undefined specialised terms that are hardly informative. After hearing his lengthy hyperbolic encomium on the beauty and wit of Baghdadi singers, the audience asks Abū al-Qāsim to favour them with some representative stories (*baʿḍ tilka al-ḥikāyāt*) as if to hold the author to the promise he made in his introduction, to be a *ḥākī* of Baghdad. Abū al-Qāsim responds at first with indignation: 'Gentlemen, are you looking for a clown? Do you want somebody to laugh at? Your friend the fool? No, sir, find someone else to laugh at!'[21]

And the same guest would reply, 'God forbid, Abū al-Qāsim! If you favoured us we would thank you, and you would be respected as a gentleman, not ordered about! If you refuse, we won't ask you for anything like it again, and you would still be an honoured guest here!'[22]

This section again raises questions of the correct definition of the word *ḥikāya*. Abū al-Qāsim seems to think that by asking him to produce *ḥikāyāt* (stories or, perhaps, imitations of the slave girls) the other guests are asking him to reduce himself to the role of popular entertainer or fool. After being mollified by the guests, what Abū al-Qāsim does produce are a series of anecdotes on slave girls in Baghdad, a section which more closely resembles a typical collection of *akhbār* (anecdotes) in the *adab* tradition than anything else in the *Ḥikāya*. These, however, are not called *akhbār*, but rather *ḥikāyāt*, the difference lying perhaps both in their imitative nature and their low implied truth value; Abū al-Qāsim's *ḥikāyāt* about the witty Baghdadi slave girls includes several short anecdotes with punchlines that belong more in a joke book than a history book. He ends with a string of stories in which listeners react so strongly to the beauty of a slave girl's song, tearing their hair and writhing on the floor, that Abū al-Qāsim's incredulous audience has to ask, 'All this just from listening to some singing?'[23]

The equation here, however, of *ḥikāya* with a type of *khabar*, confirms the impression that *Ḥikāyat Abī al-Qāsim* can be read as a single, book-length *khabar* – as if it is a section of a larger book in extreme close-up. This impression further contributes to the text's microcosmic feeling, in which a small part might represent a universe of being. That the book ends much as it begins, and is told in the iterative mode, suggests that similar stories may exist on either side. We are consequently led to speculate about Abū al-Qāsim's life outside the text.

Thus, though we may accept that he represents a microcosm, we might not be able not to think of him also as a human being. Through his poetry we may glean that he once lived in Isfahan,[24] went to Baghdad, and then left that city leaving behind everyone that he loved, but then again, he may simply be quoting poetry to that effect. The iterative mode of the narrative would impress the reader with the belief that he would habitually do the things described, surely an exhausting way of life. Does Abū al-Qāsim still do these things? Has he died? Can he die? Is he man or monster, and should we pity him? Or to put it another way, is he monster or abject hero? As Michael Bernstein describes in *Bitter Carnival*:

> The monster is monologic in his self absorption, the Abject Hero is condemned to dialogue, since his conscious is an echo chamber of incompatible desires and prohibitions, a sound box in which the voices of the monster, the contentedly successful citizen, the desperately hungry parasite, and the resigned failure exchange insults and advice with bewildering inconsistency.[25]

Even the abject hero, as described in this quotation, is in dialogue solely with himself, and so it often seems with Abū al-Qāsim. Like modern scholars, his Isfahani audience clearly want him to tell them something about Baghdad. They ask him about the identity of the Baghadadi man he describes only as *fulān* (so-and-so), about his house, about swimming and sailing in the city, and either receive the brush-off or an incomprehensible list of words in reply. His tyrannical refusal of the audience's right to laugh is only the worst example of his abuse of the guest and of the reader. He may put us in mind of a mad emperor, a Caligula ruling the feast with his words, or even Tommy DeVito of *Goodfellas* terrifying his friends with the question, 'I'm funny how,

I mean funny like I'm a clown, I amuse you? What do you mean funny, funny how? How am I funny?'

Rosenthal's *Humor in Early Islam* provides al-Tawḥīdī's definition of laughter (from a chapter on the subject in the *Muqābasāt* (Borrowed Lights):[26] 'Laughter is a power originating in between the powers of reason and animality.' It is connected with reason in that it results from astonishment and the attempt to discover 'the reason and cause of things that happen to occur'. It is connected with the life of the body in that it relates to the movement of the 'animal power', which can move inward or outward. If this power moves outward suddenly, it generates anger (if slowly, it generates joy). If it moves inward suddenly it generates fear. But if it is pulled inward and then outward 'when the two powers are pulled in different directions in search of the reason (for some phenomenon)', because someone 'decides that the reason is such-and-such, and then again, that it is not so', then 'the multiplicity of impressions [produces a] tittering in the face ... but anger then gets the upper hand over them one by one.'[27]

If anything could convince me of al-Tawḥīdī's authorship of the *Ḥikāya*, it is this quotation.[28] The complex relation of laughter, anger, and astonishment, with anger ultimately triumphing over laughter, seems to perfectly describe the dance between the quarrelsome Abū al-Qāsim and the bewildered, amused, and annoyed guests. Seen in the context of this quote, in fact, this interaction itself becomes a kind of metaphor for the struggle of the rational faculty to seek reasons beyond the capacity of language. Rosenthal considers al-Tawḥīdī's treatment of this topic remarkable especially for its including anger in its definition of laughter, and the two reactions of laughter and anger are emphatically related to one another in *Ḥikāya*.

But what of the *Ḥikāya*'s mysterious author, about whom we have said very little? Indeed, in the edition cited throughout much of this present study, he is identified as al-Tawḥīdī, and not by the name given to him at the beginning of the manuscript, Muḥammad ibn Aḥmad Abū al-Muṭahhar al-Azdī. The mystery of the author's identity has dominated scholarship on the *Ḥikāya*, as if the discovery of that identity might be the key that would unlock the secrets of this strange text with its strange protagonist. But the author's narrative voice is almost laughably reasonable and modest in comparison to his wildly overbearing protagonist, as if his presence were effaced by Abū al-Qāsim's conversation along with that of the other Isfahani guests. To return to the lines from *Bitter Carnival* above, perhaps he is the abject hero, and Abū al-Qāsim the monster. However:

> To mimic the monstrous is still to be only a mimic, and to model one's speech after the mad is still to be dependent on prior examples. But, paradoxically, to desire such a voice for oneself *is* genuinely monstrous ... So the Abject Hero is again doomed to a double existence: parodying a role that is, in reality, already his own, and imitating a state that he already inhabits.[29]

That is to say, it may be true we know little more of al-Azdī than his authorship of the *Ḥikāya*, but perhaps, in knowing him as the man behind the microcosm, we know more than enough.

With a strikingly self-referential gesture, the *Ḥikāya* addresses its own author at the end of the story, commanding him to bear witness, and, as it were, to write the story that was just told ('Welcome to a new day and a writer bearing witness! Write: In the name of God the merciful, the compassionate, Abū al-Qāsim ʿAlī ibn Muḥammad al-Tamīmī al-Baghdādī says ...')[30] Thus the world of the *Ḥikāya* encloses within its microcosmic confines even its own author, who is charged, prophet-like, with the task of conveying its message.

Philemon and Baucis

> But in gazing at such scenes, it is all in all what mood you are in; if in the Dantean, the devils will occur to you; if in that of Isaiah, the archangels.
>
> Melville, *Moby-Dick*, 'The Tail'

In Ovid's *Metamorphoses*, the gods Zeus and Hermes, disguised as mortals, are denied hospitality at every house except one. At the humble house of the aged couple Philemon and Baucis, the gods are treated with generosity and respect. As a reward, their poor dwelling is transformed into a magnificent temple, and they into its priests. Thus we learn that when the gods show up as uninvited guests, nothing in one's house remains the same.

While the gods Zeus and Hermes are still pretending to be human, and Baucis and Philemon's home remains a hovel, the narrator provides a detailed description of the meal that the poor couple serves:

[Baucis] the wood split fine, and the dry twigs, made smaller
By breaking them over the knee, and put them under
A copper kettle, and then she took the cabbage
Her man had brought from the well-watered garden,
And stripped the outer leaves off. And Philemon
Reached up, with a forked stick, for the side of bacon,
That hung below the smoky beam, and cut it [...]
With trembling hands. One table-leg was wobbly;
A piece of shell fixed that. She scoured the table,
Made level now, with a handful of green mint,
Put on the olives, black or green, and cherries
Preserved in dregs of wine, endive and radish,
And cottage cheese, and eggs, turned over lightly
In the warm ash, with shells unbroken. The dishes,
Of course, were earthenware.[31]

By comparison to Abū al-Qāsim's fantastic lists of foodstuffs, Ovid's list feels very realistic. This realism heightens the drama of the humble house's later transformation into a grand temple. As part of a poem with epic pretensions, however, the realism of this description is of a heroic breed. The *Satyrica* provides a parodic response to Ovid's realism in its description of the ramshackle house belonging to Oenothea, priestess of Priapus:

> While she plucked off a little bit of meat and was putting a pork head as old as she was back on the meat rack, she broke a putrid little stool by whose height she was increasing her own, sending her aged weight flying onto the brazier. The neck of the pot broke, extinguishing the just-reviving fire. She burned her elbow on a flaming stick, and poured the spilled ashes all over her face.[32]

This description of a failed dinner is a clear parody of Ovid's 'Philemon and Baucis'.[33] The disjunction lies between two types of realism: the almost heroically humble description in the *Metamorphoses* and another kind of realistic description, one that emphasises the messiness of life in contrast with the tidiness of the epic register. The story of 'Philemon and Baucis', though its subject begins humbly, suits the grandeur of the Olympian gods, Zeus and

Hermes. The *Satyrica* is home to a much more earthly god, Priapus, and the language there also matches the nature of the divinity concerned.[34]

Thus, not only do the gods in these works shape and change objects, but language itself as well. Priapus, a roughly-carved garden god who comically threatens vegetable thieves with his enormous phallus, is a pervasively implied presence in the *Satyrica*. In the description of the house of the priestess of Priapus quoted in part above, a gaggle of sacred geese soon crash the party, their long necks evoking Priapus' phallus. Encolpius, the hapless and impotent protagonist of the tale, bludgeons the aggressive leader of these sacred geese with a table-leg, further incurring the wrath of the garden god. He then compares himself to Heracles performing one of his twelve labours by attacking the Stymphalian birds.[35] In this episode, and throughout the *Satyrica* (a tale driven by the wrath of Priapus), the language is a suitably humorous version of the epic narratives inhabited by the Olympians (the story of Odysseus, for example, driven by the wrath of Poseidon).[36] As for Ovid's narrative of Philemon and Baucis, it contains elements of both heaven and earth, and Hermes, a liminal god of Olympus, is instrumental in the sublime transformation of Philemon and Baucis' humble abode. Like him, their house and the language of the poem make the journey between one realm and the next, beginning with a simple description of furnishings and food, and ending with their dizzying transformation into a temple of the gods.

Divine party-crashers, like prophets, disrupt the perceived reality of the earthly world with a higher truth. The word 'mimesis' or *muḥākāh* (from the same root as *ḥikāya*) is used in al-Fārābī's *The Perfect City* to describe this ability to see higher forms of knowledge in earthly imitations, or symbols, which faculty is found especially among prophets (or during dreams, themselves considered a fraction of prophecy).[37] The perfect man (of whom Abū al-Qāsim is a sort of Doppelgänger) is characterised partly by this imitative visionary faculty.[38] Language itself is a kind of imitation,[39] and like the imaginative imitations of a prophetic vision, language can represent another world. Even the language of an Arabic wine poem can evoke the drunkenness of a spiritual ecstasy, itself a form of worship.[40] Language can form a crucial part of spiritual transcendence and transformation, as does that of the Qur'an. The Qur'an, in turn, could perhaps be described as a *ḥikāya* according to al-Fārābī's definition, as it is an earthly manifestation of a divine truth that is contained in a heavenly 'guarded tablet'.[41]

But poetry, unlike prophecy, is not beholden to the truth, but rather, in its self-referential or deceitful capacities, can question reality and create a world otherwise unreal. Stefan Sperl writes that classical mimesis (a concept related to al-Fārābī's use of the word *ḥikāya*) runs counter to this mannerist style of literature, which 'presupposes an attitude to reality as much as to language: an awareness of incongruity between them'.[42] Later he adds, 'Structural limits are the very instruments of semiological mimesis, providing the constraints which delimit the possible within the combinatory scope of language. These constraints act like prisms, revealing in language a self-contained and boundless world of patterns in relation.'[43] Thus language provides its own system of physics within which to create a new world. While mimetic language may create a new world, this world is in harmony with physical reality. Manneristic language, in contrast, also creates a new world, but the effect of this creation is to question perceptions of reality and the ability of language to represent the real.[44] Sometimes in creating these new worlds, language has the ability even to transform the real, or the objects that it describes.

The ambiguity and power of language makes it the perfect tool of the trickster, whether a lowly party-crasher or the great god Hermes. In the Arabic literary tradition, poets in particular are known to 'say what they do not do'.[45] This type of speech can nevertheless have a very real effect on our perceptions of the things around us. As van Gelder writes:

> We should not be surprised that such a high proportion of court jesters known from the sources are poets … judging by the author's [Niẓāmī 'Arūḍī's][46] definition of poetry, the function of which is, as it were *qalb al-ashya'* [reversing things]: 'Poetry is that art whereby the poet arranges imaginary propositions and blends fruitful analogies, in such wise that he can make a little thing appear great and a great thing small, or cause good to appear in the garb of evil and evil in the garb of good.'[47]

Alvin Kernan uses similar language to describe the essential features of the genre of satire, in which language itself can become a kind of trickster. Satire's 'magnifying tendency' makes heroic what is trivial, and its 'diminishing tendency' involves 'the diminution of idea to thing'.[48]

Jonathan Swift's satirical rewriting of 'Baucis and Philemon' is a good example of satire that diminishes idea to thing. This poem focuses on the transformation of ordinary household objects into items in a church: for example, the chimney turns into a spire and the kettle in the fireplace becomes the church bell:

The heavy wall climbed slowly after.
The chimney widened, and grew higher,
Became a steeple with a spire.
The kettle to the top was hoist,
And there stood fastened to a joist;
But with the upside down, to show
Its inclination for below.
In vain; for a superior force
Applied at bottom, stops its course,
Doomed ever in suspense to dwell,
'Tis now no kettle, but a bell.[49]

Although the objects in the house are magnified in size, the satirical language of the poem seems to diminish the trappings of a lofty church, which now appear only the magnified furnishings of a hovel, putting on airs. Swift's satirical rewrite also focuses on Philemon and Baucis' stingy neighbours, who, though Christians, show no charity.

What effect does Abū al-Qāsim's unexpected arrival have on the house he visits and the objects and people therein? Some have seen his language as similar to Swift's in its diminishing portrayal of various pious and respectable classes.[50] Abū al-Qāsim indeed lampoons all the guests and their professions, his sarcasm doubling in bitterness when they defend one another as 'noble'. But perhaps it better fits another tendency Kernan discerns in satirical language, the 'mob tendency', which, a kind of motley stew of language, shows each ingredient in jarring juxtaposition.[51] As an example of this mode of satire, Kernan describes *The Day of the Locust*'s portrayal of Los Angeles as a clash of multiple feigned styles whose proximity to one another only magnifies their falseness. As shown in previous chapters, the *Ḥikāya* has little to say about the people or objects in the physical cities of Baghdad and Isfahan. Abū al-Qāsim's descriptions of Baghdad, Isfahan, and the many things found in both cities, overflow with disparate ingredients, and almost taunt us with the unreality of these objects cloaked in language. We are left not knowing what to believe about the people and things present at the party, nor about the absent people or things with whom they are compared.

Compare Ovid's list of foods in the poor couple's house, cited at the beginning of this section, to some of Abū al-Qāsim's descriptions of food and dinner service in the *Ḥikāya*, which, quoted in full, would occupy tens of pages:

> Among the grilled items are Kaskarī duck and Ṣarṣarī kid, and Indian fattened chicken, and Turkoman suckling lamb, round, its length and breadth the same, just plucked from the teats of its mother, cooked as hot as a nest of hornets. Fattened chicks sweeter than health, atop sugared walnuts, and rice with milk, with saffron thrown in, studded with chickpeas, and strewn with powdered sugar. Sweet bountiful Jaʿfariyya, that Baghdadi born in Byzantium. Saffron meat strips, like rods of myrtle, the table like a bride bedecked, surrounded with ornaments: blood red opposite bright yellow, deep black next to shining white, a kid as red as a rose and anemonies, *būriyya* fish as white as white Egyptian linen, a mouthful drowning in its own oil before it reaches the rice. The meat and fat of this kid builds a red and white foundation in the stomach, by which a ballista of stones (meaning a drink) hurled against it would be deflected, making no impression at all. Yes! Turtle doves in thickened cheese stew, drowned in their own oil, and pickled fish from a clay oven, fried ring doves, quail, partridges, chicks, stall-fed breasty chickens, golden skinned, silver fleshed, Indian or Barhindiyyan, thick-thighed, heavy breasted, saturated with fat, fed with barley flour. Purified oil, pressed from greases, dates, sausages, braided meat, filets, Rashidi kebabs, spiced flanks, pregnant pigeons, chicks still fed in the nest, francolins and geese just sprouting feathers, grilled sides dripping, O God! juice, the meat flowing with grease and broth,[52] and shawarma, and pomegranate vinaigrette with *sikbāj* cooked in wine vinegar, with young lamb's meat on top, and chicks just ready to fly, and water fowl, and yellow songbirds with peeled almonds inside, and Khurāsānī raisins, and Jurjānī jujube, garnished with citron leaves.[53]

This ecstatic orgy of foods begins on a previous page with Abū al-Qāsim asserting that he sees no such luxuries in Isfahan. Thus this list is presented not as a list of things found in Baghdad, but as a list of things not found in the supposedly inferior city of Isfahan (as described in Chapters 2 and 3). Abū al-Qāsim's replacement of food with words itself constitutes a kind of metamorphosis. This (pseudo?) divine party-crasher replaces a barely-described but realistic-sounding meal served in Isfahan (e.g. 'a platter of cheese ... and some pickled snacks') with an unlikely-sounding absent feast of language. In the *Metamorphoses*, because Baucis and Philemon were such good, pious people, their house was always a temple to the gods, and so the transformation enacted by their unexpected visitors perhaps served only to show things in their truest light. The transformation enacted by Abū al-Qāsim, in contrast, leaves the truth an open question.

Unlike Zeus and Hermes, Abū al-Qāsim leaves his host and the house physically much as he found them. Psychologically, however, his visitation has an impact. The room, at least, must have spun, though it never transformed into a temple. In hosting Abū al-Qāsim, it hosted a microcosm, its boundaries, like a bottle with a ship of fools inside, strangely altered by the impossibility of its contents.

Notes

1. Gabrieli, 'Sulla *Ḥikāyat*', p. 34.
2. *Ḥ*, p. 335.
3. It is interesting to note that Plato's *Symposium* similarly ends with a drunken, chaotic moment of revelry, and that Socrates and some other guests drink again the next morning, while Socrates explains that a writer of tragedy should also be able to write comedy. This seems to suggest something similar about the balance of jest with seriousness during certain appropriate times, while similarly upsetting this balance (see 'Those Camels have Passed', Chapter 2, p. 92).
4. 'That is my praise for you and for mankind | You drip of a cuckold's wet dick | And if you should feel that I've cut it too short | Just think this a garden pic-nic' (*Ḥ*, p. 349).
5. Kwei-Armah, *The Man Who Was Bojangles*.
6. Moreh, *Live Theatre*, pp. 94–101. Hämeen-Anttila, 'al-Hamadhānī', p. 85. St. Germain, *Anomalous*, p. 117.
7. *Ḥ*, p. 349.
8. Bakthin, *Rabelais*, p. 207.
9. *Ḥ*, pp. 358, 367.

10. *Ḥ*, p. 369.
11. Indeed, this is a trait attributed to medieval Baghdadis. See, for example, Bosworth, *Mediaeval Islamic Underworld*, p. 54, n. 16. Also see 'Baghdad the Party-Crasher', Chapter 2, pp. 81–2).
12. Hamori, *On the Art*, p. 56.
13. *Ḥ*, p. 390.
14. As, for example, when he farts, angering the beautiful singing girl (*Ḥ*, p. 360).
15. *Ḥ*, p. 391.
16. Hyde, *Trickster*, p. 219.
17. Hyde, *Trickster*, p. 7.
18. Lines pp. 52–62. See Vergados, *Hymn to Hermes*, p. 272.
19. Hyde, *Trickster*, p. 287.
20. Hyde, *Trickster*, p. 244.
21. *Ḥ*, p. 229.
22. *Ḥ*, p. 229.
23. *Ḥ*, p. 246.
24. 'You ask me about Isfahan (may time yet tear it down!); | the young men look like old men there, the old resemble hounds. | I left when just a child, and thus escaped those sterile grounds!' (*Ḥ*, p. 90). As mentioned above in 'Baghdad the Party-Crasher' in Chapter 2 (p. 79), this poem is elsewhere ascribed to Abū al-Faḍl al-Muẓaffar ibn Aḥmad al-Yazdī.
25. Bernstein, *Bitter Carnival*, p. 27.
26. 'Borrowed Lights' is Charles Genequand's translation of the title of this work, and refers to the practice of borrowing live coals from a neighbour in order to start a new fire, an image often used as a metaphor for literary borrowing and inspiration. Following translations of al-Ṭawḥīdī's thoughts on laughter are Rosenthal's.
27. Rosenthal, *Humor*, pp. 137–8. A variant reading of the Arabic in this last phrase would instead give us, 'The nerves hang from it one by one' (reading *wa-yaʿlaqa al-ʿaṣab* instead of *wa-yaʿlū al-ghaḍab*). See al-Tawḥīdī, *Muqābasāt*, pp. 294–5.
28. The debate about the authorship of the *Ḥikāya* and my decision not to enter fully into this debate are explained in my introduction.
29. Bernstein, *Bitter Carnival*, p. 31.
30. *Ḥ*, 390 (cited more fully above in Chapter 5, pp. 157–8). Cf. Q 3:35, 'Our Lord! we believe in what thou hast revealed, and we follow the Messenger; then write us down among those who bear witness' (trans. Yusuf Ali).

31. Ovid, *Metamorphoses* VIII: 645 ff. in Tarrant (ed.), *P. Ovidi Nasonis Metamorphoses.*
32. Translation mine. Petronius, *Satyricon* 136: 1–8.
33. Rosenmeyer provides a detailed analysis of the parodic features of this episode in 'The Unexpected Guests', pp. 403–13.
34. Similarly, the Homeric hymn to Hermes, in its playful, irreverent tone, reflects the nature of the divinity honoured by the hymn. This was brought to my attention by Thomas, 'The Lyre's Paradox in the *Homeric Hymn to Hermes*'. His commentary on the Homeric hymn to Hermes is forthcoming in the Cambridge Classical Texts series.
35. Petronius, *Satyricon* 136: 4–5. Also see Schmeling, *Commentary*, p. xlviii.
36. For an interpretation of the *Satyricon* as parody of the Odyssey, see Schmeling, *Commentary*, pp. 523–9, and Richlin, *Garden of Priapus*, p. 192, as well as her 'Sex in the *Satyricon*', p. 95. Priapus himself, as he is portrayed in the collection of humorous dedicatory poems known as the *Priapea*, seems acutely aware of his low, parodic status in comparison with the lofty Olympic gods. See, for example, poem twenty in which he compares his giant phallus to Jove's lightning bolts, Neptune's trident, and the other gods' phallic-shaped weapons. Also see poems 36 and 75 in Hooper, *Priapus Poems*, which include his (loose) translations facing the Latin.
37. Al-Fārābī, *al-Madīna*, pp. 210–27, 416. Also see Walzer, 'Al-Fārābī's Theory', pp. 211–16.
38. Al-Fārābī, *al-Madīna*, p. 439.
39. Pellat raises this point in discussing the philologists al-Zamaksharī's (d. 1143) *Asās al-balāgha* (Foundation of Eloquence) in '*Ḥikāya*'. Also see Plotinus in Halliwell, *Aesthetics*, p. 315.
40. As Montgomery writes, 'The Symposium was the venue at which God was worshipped and the divine light was celebrated in drinking and in songs.' Wine itself, he adds, could serve as a creator or a door to a new world ('Justified Sinner?', pp. 127–30. This article uses the wine poetry of Abū Nuwās as a source of information about theological and philosophical matters). Also see Hamori's *On the Art*, p. 56, where he describes how drunkenness can be a metaphor for a spiritual unawareness of time and the mundane.
41. Q 85:22.
42. Sperl, *Mannerism*, p. 164.
43. Sperl, *Mannerism*, p. 180. Also see Heinrichs on *takhyīl*: 'a shift of the poet's attention from the level of reality, that is, the level of the objects depicted, to

the level of imagery which is erected on a line parallel with reality, but which nonetheless allows the poet ... to create a phantastic microcosm at will' (*Literary Theory*, p. 26). *Takhyīl* has been translated as *phantasia;* the definition of this word, its Neoplatonic import, and its classical Greek heritage is discussed in Sheppard's introduction to *Takhyīl.*

44. Robert Irwin expresses a similar idea in summarising al-Jurjānī's *Asrār al-balāgha:* 'Jurjani argued that language was a convention and that words, and indeed metaphors and similes, had no independent meaning, but depended on their placement in a linguistic whole' (*Night and Horses*, p. 216).
45. Q 26:226.
46. Niẓāmī ʿArūḍī was a twelfth-century Persian writer from Samarqand, best known for his work *Chāhār maqālih* (Four Discourses), from which this quote is derived.
47. Van Gelder, quoting Edward G. Browne's translation in 'Fools and Rogues', p. 37.
48. Kernan, *Plot of Satire*, pp. 36, 52. Ibn Qutayba uses similar language in criticising al-Jāḥiẓ's writing about 'low' topics, saying that he 'makes trifles great and great things trifles' (from Mez, *Renaissance of Islam*, p. 204).
49. Swift, *Poems*, pp. 92–3.
50. See, for example, Balda, 'Marginalité', pp. 386, 393.
51. The root of the word 'satire', *lanx satura*, is in fact a stew of motley and disparate ingredients.
52. For *jūdhāba* see Perry, *Medieval Arab Cookery*, pp. 81–82.
53. *Ḥ*, pp. 157–60. Also cited in part in 'Baghdad the Party-Crasher' in Chapter 2, p. 84.

Bibliography

ʿAbd al-Mawlā, Muḥammad Aḥmad, *al-ʿAyyārūn wa-l-shuṭṭār al-Baghdādiyya fī al-tārīkh al-ʿAbbāsī* (Alexandria: Muʾassasat Shabāb al-Jāmiʿah, 1987).

Ahsan, Muhammad, *Social Life under the Abbasids* (London: Longman Group, 1979).

al-Ājurrī, Muḥammad ibn al-Ḥusayn, *Taḥrīm al-nard wa-l-shaṭranj wa-l-malāhī* (Beirut: Dār al-Kutub al-ʿIlmiyya, 1988).

Ali, Samer, *Arabic Literary Salons in the Islamic Middle Ages: Poetry, Public Performance, and the Presentation of the Past* (Notre Dame: University of Notre Dame Press, 2010).

Allen, Roger, *The Arabic Literary Heritage* (Cambridge: Cambridge University Press, 1998).

Anderson, Graham, *The Second Sophistic* (London: Routledge, 1993).

Antoon, Sinan, *The Poetics of the Obscene in Premodern Arabic Poetry: Ibn al-Ḥajjāj and Sukhf* (New York: Palgrave Macmillan, 2014).

Apuleius, *Metamorphoses*, in R. Helm (ed.), *Apuleius I:* Metamorphoseon *Libri XI* (Leipzig: Teubner, 1955).

Aristotle, *Poetics*, in Stephen Halliwell (trans.), *The Poetics of Aristotle* (London: Duckworth, 1987).

Auerbach, Erich, *Mimesis: The Representation of Reality in Western Literature* (Berne: A. Francke, 1946).

Awn, Peter, *Satan's Tragedy and Redemption: Iblīs in Sufi Psychology* (Leiden: Brill, 1983).

Azarnoosh, Azartash, 'Abū al-Muṭahhar al-Azdī', trans. Suheyl Umar, in *Encyclopaedia Islamica* (Brill Online, 2012).

Al-Azdī, Muḥammad ibn Aḥmad Abū al-Muṭahhar, *Ḥikāyat Abī al-Qāsim al-Baghdādī*, Oriental and India Office Collections, British Library, London, ADD 19, 913.

——— *Ḥikāyat Abī al-Qāsim al-Baghdādī / Abulḳâsim: ein bagdâder Sittenbild*, ed. Adam Mez (Heidelberg: Carl Winter, 1902).

——— *Ḥikāyat Abī al-Qāsim al-Baghdādī*, published as Abū Ḥayyān al-Tawḥīdī [attributed], *al-Risālat al-Baghdādiyya*, ed. ʿAbbūd al-Shāljī (Koln: Manshūrāt al-Jamal, 1980).

——— *24 heures de la vie d'une canaille*, trans. R. C. Khawam (Paris: Phébus, 1998).

Baker, Nicholson, *Mezzanine* (New York: Vintage Books, 1988).

Bakhtin, Mikhail, *The Dialogic Imagination*, trans. Caryl Emerson and Michael Holquist (Austin: University of Texas Press, 1981).

——— *Rabelais and His World*, trans. Hélène Iswolsky (Bloomington: Indiana University Press, 1984).

Balda-Tillier, Monica, 'Marginalité et eloquence contestatoire: le personage d'Abū l-Qāsim dans la *Ḥikāyat Abī l-Qāsim* d'Abū al-Muṭahhar al-Azdī', in *Identidades Marginales*, ed. Cristina de la Puente (Madrid: CSIC, 2003): 371–93.

——— '*ʿUdhrī* Love and *Mujūn: Opposites and Parallels*', in *The Rude, the Bad and the Bawdy: Essays in Honour of Professor Geert Jan van Gelder*, ed. Adam Talib, Marlé Hammond, and Arie Schippers (Gibb Memorial Trust, 2014): 123–40.

Barletta, Vincent, *Covert Gestures: Crypto-Islamic Literature as Cultural Practice in Early Modern Spain* (Minneapolis: University of Minnesota Press, 2005).

Barton, Carlin, *Sorrows of the Ancient Romans* (Princeton: Princeton University Press, 1992).

Beaumont, Daniel, 'A Mighty and Never Ending Affair: Comic Anecdote and Story in Medieval Arabic Literature', *Journal of Arabic Literature* 24 (1999): 139–59.

Bernstein, Michael, *Bitter Carnival: Ressentiment and the Abject Hero* (Princeton: Princeton University Press, 1992).

Bigger, Andreas, 'Mez, Adam', in *Dictionnaire historique de la Suisse*, accessed 17 April 2012. http://www.hls-dhs-dss.ch/textes/f/F44647.php

Bodman, Whitney S., 'Stalking Iblīs: In Search of an Islamic Theodicy', in *Myths, Historical Archetypes, and Symbolic Figures in Arabic Literature*, ed. Angelika Neuwirth et al. (Stuttgart: Steiner, 1999): 247–70.

Borges, Jorge Luis, *Collected Fictions*, trans. Andrew Hurley (New York: Penguin Books, 1998).

Bosworth, Clifford Edmund, *The Mediaeval Islamic Underworld: The Banu Sasan in Arabic Society and Literature* (Leiden: Brill, 1976).

Boyle, John Andrew, 'The Death of the Last ʿAbbāsid Caliph: A Contemporary Muslim Account', *Journal of Semitic Studies* 6/2 (1961): 145–6.

Branham, R. Bracht, 'Defacing the Currency: Diogenes' Rhetoric and the *Invention* of Cynicism', in *The Cynics*, ed. R. Bracht Branham and Marie-Odile Goulet-Cazé (Berkeley: University of California Press, 1996): 81–104.

Breslin, Andy, 'The Number of Possible Different Games of Chess', in *Andy Rants*, accessed 11 April 2012. http://andyrantsandraves.blogspot.com/2009/10/number-of-possible-different-games-of.html

Brockelmann, Carl, 'Muḥammad ibn Aḥmad Abulmuṭahhar Alazdi ...' *Literarisches Centralblatt für Deutschland* 53/47 (22 November 1902): 1568–9.

Brown, Norman, *Apocalypse and/or Metamorphosis* (Berkeley: University of California Press, 1991).

Ceccarelli, Paola, Athenaeus: A Walking Library? On the Nature and Purpose of the Deipnosophistae. Paper presented to the Department of Classics and Ancient History, University of Manchester, 14 September 2013.

Caillois, Roger, *Man, Play and Games*, trans. Meyer Barash (Chicago: University of Chicago Press, 2001).

Carroll, Lewis, *Through the Looking Glass*, in *The Annotated Alice*, ed. Martin Gardner (New York: W. W. Norton, 2000).

Chessbase News, accessed 15 April 2012. http://www.chessbase.com/newsdetail.asp?newsid=153

Christian, Brian, *The Most Human Human: What Talking with Computers Teaches us about what it Means to be Alive* (New York: Doubleday, 2011).

Conger, George, *Theories of Macrocosms and Microcosms in the History of Philosophy* (New York: Columbia University Press, 1922).

Connors, Catherine, *Petronius the Poet: Verse and Literary Tradition in the Satyricon* (Cambridge: Cambridge University Press, 1998).

Cooperson, Michael, 'Baghdad in Rhetoric and Narrative', *Muqarnas* 13 (1996): 99–113.

——— 'Images without Illustrations: The Visual Imagination in Classical Arabic Biography', in *Islamic Art and Literature*, ed. Oleg Grabar and Cynthia Robinson (Princeton: Markus Wiener Publishers, 2001): 7–20.

Corbett, Philip, *The Scurra* (Edinburgh: Scottish Academic Press, 1986).

Corner, Sean, 'The Politics of the Parasite (Part One)', *Phoenix* 67/1–2 (2013): 43–80.

——— 'The Politics of the Parasite (Part Two)', *Phoenix* 67/3–4 (2013): 223–36.

Crone, Patricia, 'Imperial Trauma: The Case of the Arabs', *Common Knowledge* 12 (2006): 107–16.

——— 'Post-Colonialism in 10th century Islam', *Der Islam* 83 (2006): 2–38.

Davidson, James, *Courtesans and Fishcakes: The Consuming Passions of Classical Athens* (New York: HarperCollins, 1997).

——— 'Pleasure and Pedantry in Athenaeus', in *Athenaeus and his World*, ed. David Braund and John Wilkins (Exeter: University of Exeter Press, 2000): 292–303.

De Goeje, M. J., 'Abulkasim von Muḥammad ibn Ahmad, hrsg. von Mez', *Göttingische gelehrte Anzeigen* 164/IX (1902): 723–36. (Review.)

Dhū al-Nūn Ṭāhā, ʿAbd al-Wāḥid, 'Mujtamaʿ Baghdād min khilāl Ḥikāyat Abī al-Qāsim al-Baghdādī', *al-Mawrid* 3 (1974): 14–25.

Dozy, Reinhart Peter Anne, *Supplément aux dictionnaires arabes* (Leiden: Brill, 1927).

Elliott, Robert, *The Power of Satire* (Princeton: Princeton University Press, 1960).

Al-Fārābī, Abū Naṣr, *Mabādiʾ ārāʾ ahl al-madīna al-fāḍila: A Revised Text with Introduction, Translation, and Commentary by Richard Walzer* (Oxford: Clarendon Press, 1985).

Farrin, Richard, *Abundance from the Desert: Classical Arabic Poetry* (Syracuse: Syracuse University Press, 2011).

Freytag, G. W., *Arabum Proverbia* (Bonnae: A. Marcum, 1838).

Gabrieli, Francesco, 'Abū l-Muṭahhar al-Azdī', in *Encyclopaedia of Islam* XII, 2nd edn (Leiden: Brill, 1980): 31.

——— 'Sulla *Ḥikāyat Abī l-Qāsim* di Abū l-Muṭahhar al-Azdī', *Rivista degli Studi Orientali* 20 (1942): 33–45.

Genette, Gérard, *Narrative Discourse: An Essay in Method*, trans. Jane E. Lewin (Ithaca: Cornell University Press, 1980).

Giles, Ryan, *The Laughter of the Saints: Parodies of Holiness in Late Medieval and Renaissance Spain* (Toronto: University of Toronto Press, 2009).

Goitein, S. D., *A Mediterranean Society, Volume IV: Daily Life* (Berkeley: University of California Press, 1983).

Goskar, Tehmina, 'Material Worlds: The Shared Cultures of Southern Italy and its Mediterranean Neighbours in the Tenth to Twelfth Centuries', *al-Masaq: Islam and the Medieval Mediterranean* 23/3 (2011): 189–204.

Gowers, Emily, *The Loaded Table: Representations of Food in Roman Literature* (Oxford: Oxford University Press, 1993).

Gutas, Dimitri, 'Plato's *Symposion* in the Arabic Tradition', in *Greek Philosophers in the Arabic Tradition* (Burlington: Ashgate Publishing, 2000): 36–60.

——— 'Sayings by Diogenes Preserved in Arabic', in *Le Cynisme ancien et se prolongements*, ed. Marie-Odile Goulet-Caze and Richard Goulet (Paris: Presses Universitaires de France, 1993): 475–518.

Haddawy, Husain (trans.), *The Arabian Nights* (New York: Norton, 1990).

Al-Ḥallāj, Ḥusayn ibn Manṣūr, *Kitāb al-ṭawāsīn*, ed. Louis Massignon (Paris: Paul Geuthner, 1913).

Halliwell, Stephen, *The Aesthetics of Mimesis: Ancient Texts and Modern Problems* (Princeton: Princeton University Press, 2002).

Al-Hamadhānī, Badīʿ al-Zamān, *Maqāmāt* (Beirut: Dār al-Mashriq, 1968).

——— *Maqāmāt*, trans. W. J. Prendergast (London: Curzon Press, 1973).

Hämeen-Anttila, Jaakko, 'al-Azdī, Abū al-Muṭahhar', in *Encyclopaedia of Islam* I, 3rd edn (Leiden: Brill, 2010).

——— 'al-Hamadhānī and the Early History of the *Maqāma*', in *Philosophy and Arts in the Islamic World*, ed. U. Vermeulen and D. de Smet (Leuven: Peeters Press, 1998): 83–96.

——— *Maqama: A History of a Genre* (Wiesbaden: Harrassowitz, 2002).

Hamori, Andras, *On the Art of Medieval Arabic Literature* (Princeton: Princeton University Press, 1974).

Heinrichs, Wolfhart, 'Literary Theory: The Problem of its Efficiency', in *Arabic Poetry: Theory & Development*, ed. G. E. von Grunebaum (Wiesbaden: Harrassowitz, 1973): 19–69.

——— 'Rose vs. Narcissus, Observations of the Arabic Literary Debate', in *Dispute Poems and Dialogues in the Ancient and Mediaeval Near East* (Leuven: Peeters Press, 1991): 179–189.

Hooper, Richard W. (trans.), *The Priapus Poems: Erotic Epigrams from Ancient Rome* (Chicago: University of Illinois Press, 1999).

Horovitz, Josef, 'Abu 'l-Ḳāsim', in *The Encyclopaedia of Islam* I, 2nd edn (Leiden: Brill, 1960–2009): 133.

——— *Spuren griechisher Mimen im Orient* (Berlin: Mayer and Muller, 1905).

Al-Ḥusayn, Quṣayy, 'Muqāḍāh li-l-tazyīf wa-l-tazwīr fī al-turāth', *Al-Hayat*, 12 August 1997, no. 12944.

Hyde, Lewis, *Trickster Makes this World: Mischief, Myth, and Art* (New York: North Point Press, 1998).

Ibn Abī Ḥajala, *Unmūdhaj al-qitāl fī naql al-ʿawāl*, ed. Zuhayr Aḥmad al-Qaysī (Baghdad, 1980).

Ibn Buṭlān, al-Mukhtār ibn al-Ḥasan, *Daʿwat al-aṭibbāʾ*, ed. ʿIzzat ʿUmar (Damascus: Dār al-Fikr, 2003).

Ibn al-Nadīm, Muḥammad ibn Ishāq, *Kitāb al-fihrist*, 2nd edn, ed. Reza Tajaddod (Tehran: al-Bank al-Bazargani, c. 1970).

Ikhwān al-Ṣafāʾ, *Rasāʾil ikhwān al-ṣafāʾ* III (Beirut: Dār Beirut, 1957).

Irwin, Robert, *Night and Horses and the Desert: An Anthology of Classical Arabic Literature* (New York: Anchor Books, 1999).

Al-Jābī, Bassām ʿAbd al-Wahhāb, *Akhbār al-ḥamīr fi al-adab al-ʿarabī* (Beirut: Dār ibn Ḥazm, 2002).

Al-Jāḥiẓ, Abū ʿUthmān ʿAmr ibn Baḥr, *Al-Bayān wa-l-tabyīn*, ed. ʿAbd al-Salām Muḥammad Hārūn (Cairo: Maktabat al-Khānjī, 1968).

——— *Kitāb al-bukhalāʾ*, ed. Ṭāhā al-Ḥājirī (Cairo: Dār al-Maʿārif, 1963).

——— *The Book of Misers*, trans. R. B. Serjeant (Reading: Garnet Publishing, 1997).

——— *Kitāb al-Ḥayawān*, ed. ʿAbd al-Salām Muḥammad Hārūn (Cairo: Muṣṭafā al-Bābī al-Ḥalabī, 1966).

Jeanneret, Michel, *A Feast of Words: Banquets and Table Talk in the Renaissance*, trans. Jeremy Whiteley and Emma Hughes (Chicago: University of Chicago Press, 1991).

Al-Jīlī, ʿAbd al-Karīm, *al-Insān al-kāmil fi maʿrifat al-awāʾil wa-l-awākhir* (Cairo: Maktabat al-Thaqāfat al-Dīniyya, 2004).

——— *Universal Man*, trans. Titus Burckhardt (Paris: Dervy-Livres, 1983).

Jung, Carl, 'On the Psychology of the Trickster-Figure', in *The Archetypes and the Collective Unconscious* (Princeton: Princeton University Press, 1969): 255–74.

Kamaly, Hossein, 'Isfahan: Medieval Period', in *Encyclopaedia Iranica* XIII (London: Routledge, 2006): 641–50.

Karamustafa, Ahmet, *God's Unruly Friends: Dervish Groups in the Islamic Later Middle Period* (Salt Lake City: University of Utah Press, 1994).

——— *Sufism: The Formative Period* (Edinburgh: Edinburgh University Press, 2007).

Al-Kātib al-Isfahānī, *Kharīdāt al-qasr, qism fi dhikr fuḍalāʾ ahl Isfāhān*, ed. ʿAdnan Muhammad Tuʿmah (Tehran, 1999).

Kennedy, Philip, 'The *Maqāmāt* as a Nexus of Interests', in *Writing and Representation in Medieval Islam*, ed. Julia Bray (London: Routledge, 2006): 153–214.

Kernan, Alvin, *The Plot of Satire* (New Haven: Yale University Press, 1965).

Kahn, Muhsin (trans.), *The Noble Quran: English Translations of the Meaning and Commentary* (Madinah: King Fahd Quran Printing Complex, n.d.)

Al-Khaṭīb al-Baghdādī, Abū Bakr Aḥmad ibn ʿAlī, *al-Tatfīl wa-ḥikāyāt al-ṭufayliyyīn wa- akhbāruhum wa-nawādir kalāmihim wa-l-shʿāruhum*, ed. Bassām ʿAbd al-Wahhāb al-Jābī (Beirut: Dār Ibn Ḥazm, 1999).

——— *The Art of Party-Crashing*, trans. Emily Selove (Syracuse: Syracuse University Press, 2012).

Kilito, Abdelfattah, *The Author and His Doubles: Essays on Classical Arabic Culture*, trans. Michael Cooperson (Syracuse: Syracuse University Press, 2001).

——— *Les Séances* (Paris: Sindbad, 1983).

Kinberg, Leah, 'Ambiguous', in *Encyclopaedia of the Qur'an* I, ed. Jane Dammen McAuliffe et al. (Leiden: Brill, 2001): 70–7.

Kinney, Daniel, 'Heirs of the Dog: Cynic Selfhood in Medieval and Renaissance Culture', in *The Cynics*, ed. R. Bracht Branham and Marie-Odile Goulet-Cazé (Berkeley: University of California Press, 1996): 294–328.

Kraemer, Joel, *Humanism in the Renaissance of Islam: The Cultural Revival during the Buyid Age* (Leiden: Brill, 1986).

Krueger, Derek, 'The Bawdy and Society', in *The Cynics: The Cynic Movement in Antiquity and its Legacy*, ed. R. Bracht Branham and Marie-Odile Goulet-Cazé (Berkeley: University of California Press, 1996): 222–39.

Kurke, Leslie, *Coins, Bodies, Games, and Gold: The Politics of Meaning in Ancient Greece* (Princeton: Princeton University Press: 1999).

Kwei-Armah, Kwame, *The Man Who Was Bojangles*, BBC Radio 4, 23 August 2008.

Lambton, Ann, 'Iṣfahān', in *Encyclopaedia of Islam* IV, 2nd edn (Leiden: Brill, 1978): 105–97.

Lancioni, Giuliano, 'Gabrieli, Francesco', in *Encyclopaedia Iranica* (London: Routledge, 2000): 240–1.

Lane, Edward William, *Arabic-English Lexicon* (Beirut: Librairie du Liban, 1968).

Leder, Stefan, 'Prosa-Dichtung in der aḫbār Überlieferung Narrative', *Der Islam* 64 (1987): 6–41.

Le Strange, Guy, *Baghdad during the Abbasid Caliphate* (Oxford: Clarendon Press, 1900).

Lucian, *De Parasito*, in A. M. Harmon (trans.), *Lucian* III (Cambridge, MA: Harvard University Press, 1969).

Lyons, Malcolm C., *The Man of Wiles in Popular Arabic Literature* (Edinburgh: Edinburgh University Press, 2012).

Macdonald, Duncan Black, 'Ḥikāya', in *Encyclopaedia of Islam* III (Leiden: Brill, 1913–36): 303–5.

Magill, Frank, *Dictionary of World Biography: The Ancient World* I (Chicago: Fitzroy Dearborn, 1998).

Mallette, Karla, 'Boustrophedon: Towards a Literary Theory of the Mediterranean', in *A Sea of Languages: Rethinking the Arabic Role in Medieval Literary History*, ed. Suzanne Conklin Akbari and Karla Mallette (Toronto: University of Toronto Press, 2013): 254–66.

Marquet, Yves, 'Ikhwān al-Ṣafā'', in *The Encyclopaedia of Islam*, 2nd edn (Leiden: Brill, 1960–2009).

Marzolph, Ulrich, '"Focusees" of Jocular Fiction', in *Story-telling in the Framework of Non-Fictional Arabic Literature*, ed. Stefan Leder (Wiesbaden: Harrassowitz, 1998): 118–31.

Massignon, Louis, *Recueil de textes inédits concernant l'histoire de la mystique en pays d'Islam* (Paris: Paul Geuthner, 1929).

Al-Masʿūdī, Abū al-Ḥasan ʿAlī, *Murūj al-dhahab wa-maʿādin al-jawhar* VI, ed. Charles Pellat (Beirut: Universite Libanaise, 1966–79).

Matz, Aaron, *Satire in an Age of Realism* (Cambridge: Cambridge University Press, 2010).

Maydānī, Aḥmad ibn Muḥammad, *Majmaʿ al-amthāl* (Baghdad: Maktabat al-Muthannā, 1964).

Meisami, Julie Scott, 'Arabic *Mujūn* Poetry: The Literary Dimension', in *Verse and the Fair Sex*, ed. Frederick de Jong (Utrecht: M. Th. Houtsma Stichting, 1993): 8–30.

Melberg, Arne, *Theories of Mimesis* (Cambridge: Cambridge University Press, 1995).

Melville, Herman, *Moby-Dick* (New York: Tor Books, 1996).

Mez, Adam, *The Renaissance of Islam* (London: Luzac and Co., 1937).

Miller, Jeannie, More Than the Sum of Its Parts: Animal Categories and Accretive Logic in Volume One of al-Jāḥiẓ's *Kitāb al-Ḥayawān*. Ph.D. dissertation, New York University, 2013.

Milne, A. A., *Winnie-the-Pooh* (New York: E. P. Dutton, reprinted 1961).

Moles, John L., 'Cynic Cosmopolitanism', in *The Cynics*, ed. R. Bracht Branham and Marie-Odile Goulet-Cazé (Berkeley: University of California Press, 1996): 105–120.

Monroe, James T., *The Art of Badīʿ az-Zamān al-Hamadhānī as Picaresque Narrative* (Lebanon: American University of Beirut, 1983).

Montgomery, James, 'Abū Nuwās, the Justified Sinner?', *Abbasid Studies III: Occasional Papers of the School of ʿAbbasid Studies. Oriens* 38 (2010): 1–90.

——— *al-Jāḥiẓ: In Praise of Books* (Edinburgh: Edinburgh University Press, 2013).

——— 'Islamic Crosspollinations', in *Islamic Crosspollinations* (E. J. W. Gibb Memorial Trust, 2007).

Moreh, Shmuel, *Live Theatre and Dramatic Literature in the Medieval Arabic World* (Edinburgh: Edinburgh University Press, 1992).

Morewedge, Parviz, 'The Neoplatonic Structure of Some Islamic Mystical Doctrines', in *Neoplatonism and Islamic Thought* (Albany: SUNY Press, 1992): 51–76.

Morley, F. V., *My One Contribution to Chess* (New York: B. W. Huebsch, 1945).

Mubārak, Zakī, *al-Nathr al-fannī fī al-qarn al-rābiʿ* (Cairo: Dār al-Kātib al-ʿArabī, 1931).

——— *La Prose Arabe au IVe siècle de l'Hégire (Xe siècle)* (Paris: Librairie Orientale & Américaine Maisonneuve, 1931).

Murray, H. J. R., *A History of Chess* (Northampton, MA: Benjamin Press, 1986).

Mustawfī, Ḥamd-Allāh, *The Geographical Part of the Nuzhat al-Qulūb*, trans. Guy le Strange (Leiden: Brill, 1919).

Al-Nafzāwī, al-Shaykh, *al-Rawḍ al-ʿāṭir fī nuzhat al-khāṭir*, ed. Jamāl Jumʿa (London: Riyāḍ al-Rayyis li-l-Kutub wa-l-Nashr, 1993).

Nasr, Seyyed Hossein, *An Introduction to Islamic Cosmological Doctrines* (Albany, NY: SUNY Press, 1993).

Nicholson, Reynold, *Studies in Islamic Mysticism* (Cambridge: Cambridge University Press, 1921).

——— *A Literary History of the Arabs* (New York: Charles Scribner's Sons, 1907).

Al-Nīsābūrī, al-Ḥasan ibn Muḥammad, *ʿUqalāʾ al-majānīn*, ed. ʿUmar al-Asʿad (Beirut: Dār al-Nafāʾis, 1987).

Ovid, *Metamorphoses*, in R. J. Tarrant (ed.), *P. Ovidi Nasonis Metamorphoses* (New York: Oxford University Press, 2004).

Paulson, Ronald, *The Fictions of Satire* (Baltimore: Johns Hopkins University Press, 1967).

Pellat, Charles, 'Ḥikāya', in *Encyclopaedia of Islam* III, 2nd edn (Leiden: Brill, 1966): 367–72.

Perry, Charles (trans.), *A Baghdad Cookery Book* (Trowbridge: Cromwell Press, 2005).

Petronius, *Satyricon*, in Konrad Mueller (ed.), *Petronius:* Satyricon *Reliquiae* (Leipzig: Teubner, 1995).

Pickthall, Marmaduke (trans.) *The Meaning of the Glorious Koran: An Explanatory Translation* (London: David Cambell, n.d.).

Pileggi, Nicholas, *Goodfellas*, VHS, directed Martin Scorsese (Warner Bros, 2009).

Plato, *Symposium*, trans. Benjamin Jowett (New York: Modern Library, 1996).

Preston, Theodore, *Makamat; or, Rhetorical anecdotes of Al Hariri of Basra* (London: Cambridge University Press, 1850).

Al-Qifṭī, *Ikhbār al-ʿulamāʾ* (Cairo: Dār al-Kutub, 1926).

Ramaḍān, Yūnus, *Bughyat al-ṭālib fī maʿrifat ʿAlī ibn Abī Ṭālib* (Beirut: Muʾassasat al-Aʿlamī lil-Maṭbūʿāt, 1994).

Al-Rāwī, ʿAbd al-Laṭīf, 'A-hiya al-Risālat al-Baghdādiyya am Ḥikāyat Abī al-Qāsim al-Baghdādī?', *Majallat al-Makhṭūṭāt al-ʿArabiyya* 34/1–2 (1990).

Richlin, Amy, *The Garden of Priapus: Sexuality and Aggression in Roman Humor*, 2nd edn (New York, Oxford: Oxford University Press, 1992).

——— 'Sex in the Satyrica: Outlaws in Literatureland', in *Petronius: A Handbook*, ed. Jonathan Prag and Ian Repath (Chichester: Blackwell, 2009).

Robson, James, 'A Chess *Maqāma* in the John Rylands Library', *Bulletin of the John Rylands Library* 36/1 (1953).

Romm, James, 'Dog Heads and Noble Savages: Cynicism before the Cynics?', in *The Cynics*, ed. R. Bracht Branham and Marie-Odile Goulet-Cazé (Berkeley: University of California Press, 1996): 121–135.

Rosenmeyer, Patricia, 'The Unexpected Guests: Patterns of Xenia in Callimachus' "Victoria Berenices" and Petronius' *Satyricon*', *The Classical Quarterly* 41/ 2 (1991): 403–13.

Rosenthal, Franz, 'Fiction and Reality: Sources for the Role of Sex in Medieval Muslim Society', in *Muslim Intellectual and Social History: A Collection of Essays* (Aldershot: Ashgate, 1990).

——— *Gambling in Islam* (Leiden: Brill, 1975).

——— *A History of Muslim Historiography* (Leiden: Brill, 1952).

Rowson, Everett, 'The Effeminates of Early Medina', *Journal of the American Oriental Society* 111/4 (1991).

——— Review of *Live Theatre and Dramatic Literature in the Medieval Arabic World* by Shmuel Moreh, in *Journal of the American Oriental Society* 114/3 (1994): 466–8.

——— '*Mujūn*', in *Encyclopedia of Arabic Literature* II (London: Routledge, 1998): 546–8.

St. Germain, Mary, '*al-Azdī's* Abī al-Qāsim al-Baghdādī: Placing an Anomalous Text within the Literary Developments of its Time'. Ph.D. dissertation, University of Washington, 2006.

Al-Ṣāwī, ʿAbd Allāh, *Sharḥ Maqṣūrat Ibn Durayd* (Cairo: Maṭbaʿat al-Ṣāwī, 1951).

Schmeling, Gareth, *A Commentary on the* Satyrica *of Petronius* (Oxford: Oxford University Press, 2011).

Selove, Emily, Abū l-Qāsim, the Man behind the Mouth. Paper presented to the Department of Oriental Studies, University of Oxford, 2009.

——— 'Crashing the Text: Speaking of Eating in *Hikayat Abi al-Qasim*', in *Marginalia* 13 (Cambridge Medieval Reading Group, 2011). http://www.marginalia.co.uk/journal/11taste

——— 'Making Men and Women: Arabic Commentaries on the Gynaecological Hippocratic Aphorisms in Context', *Annales Islamologiques* 48 (2015).

——— '*Mujūn* is a Crazy Game: The Chess Game in *Ḥikāyat Abī al-Qāsim*', in *The Rude, the Bad, and the Bawdy: Essays in Honour of Geert Jan van Gelder*, ed. Adam Talib, Marlé Hammond, and Arie Schippers (Gibb Memorial Trust, 2014): 141–59.

——— The Party of Abū l-Qāsim. Paper presented at the annual meeting for the School of Abbasid Studies, Leuven, Belgium, 9 July 2010.

——— 'Who Invited the Microcosm?', *Abbasid Studies: Occasional Papers of the School of Abbasid Studies* (Gibb Memorial Trust, 2013): 76–97.

Serjeant, Robert Bertram, *Islamic Textiles: Material for a History up to the Mongol Conquest* (Beirut: Librairie du Liban, 1972).

Sheppard, Anne, 'Preface', in *Takhyīl: The Imaginary in Classical Arabic Poetics*, ed. Geert Jan van Gelder and Marlé Hammond (Gibb Memorial Trust, 2008).

Socin, Albert, *Arabische Sprichwörter und Redensarten* (Tübingen: Heinrich Laupp, 1878).

Sperl, Stefan, *Mannerism in Arabic Poetry: A Structural Analysis of Selected Texts* (Cambridge: Cambridge University Press, 1989).

Steingass, Francis Joseph, *A Comprehensive Persian-English Dictionary* (London: Boston: Routledge & Keegan Paul, 1977).

Stephen, Leslie and Sidney Lee (eds), *Dictionary of National Biography* (London: Smith, Elder and Co., 1899).

Stewart, Devin, 'Professional Literary Mendicancy in the Letters and *Maqāmāt* of Badīʿ al-Zamān al-Hamadhānī', in *Writers and Rulers: Perspectives on their Relationship from Abbasid to Safavid Times*, ed. Beatrice Gruendler and Louise Marlowe (Wiesbaden: Reichert Verlag, 2004): 39–47.

Stewart, Susan, *Nonsense: Aspects of Intertextuality in Folklore and Literature* (Baltimore: Johns Hopkins University Press, 1979).

Strogatz, Steven, 'Math and the City', *The New York Times*, 19 May 2009.

Swift, Jonathan, *The Poems of Jonathan Swift*, 2nd edn, ed. Harold Williams (Oxford: Oxford University Press, 1958).

Szombathy, Zoltan, *Mujūn: Libertinism in Medieval Muslim Society and Literature* (Gibb Memorial Trust, 2013).

Al-Ṭarābulusī, Ibrāhīm al-Aḥdab, *Kashf al-maʿānī wa-l-bayān ʿan Rasāʾil Badīʿ al-Zamān* (Beirut: al-Maṭbaʿat al-Kāthūlīkiyya, 1890).

Tarapore, J. C. (trans.), *The Explanation of Chatrang* (Bombay: Parsee Punchayet, 1932).

Al-Tawḥīdī, Abū Ḥayyān ʿAlī ibn Muḥammad, *Akhlāq al-wazīrayn* (Beirut: Dār al-Kutub al-ʿIlmiyya, 1997).

——— *al-Imtāʿ wa-l-mu'ānasa*, ed. Aḥmad Amīn and Aḥmad al-Zayn (Cairo: Lajnat al-Ta'līf wa-l-Tarjama wa-l-Nashr, 1939).

——— *Muqābasāt*, ed. Muḥammad Tawfīq Ḥusayn (Baghdad: Maṭbaʿat al-Irshād 1970).

——— *al-Risālat al-Baghdādiyya*. See: al-Azdī.

Thomas, Oliver, 'The Lyre's Paradox in the *Homeric Hymn to Hermes*'. Paper presented to the Department of Classics and Ancient History, University of Manchester, 2013.

Toone, William, *The Chronological Historian* (London: Longman Group, 1828).

Turner, Bryan, *The Body and Society: Explorations in Social Theory* (London: Sage, 1996).

Turner, Victor, *The Ritual Process: Structure and Anti-Structure* (Ithaca: Cornell University Press, 1969).

Tylawsky, Elizabeth, *Saturio's Inheritance: The Greek Ancestry of the Roman Comic Parasite* (New York: Peter Lang, 2002).

Van Gelder, Geert Jan, 'Amphigory and Other Nonsense in Classical Arabic Literature', in *Ruse and Wit: The Humorous in Arabic, Persian, and Turkish Narrative* ed. Dominic Parviz Brookshaw (Boston, MA: Ilex Foundation; Ilex Foundation Series, vol. 8, Washington, DC: Center for Hellenic Studies, 2012), pp. 7–32.

——— *The Bad and the Ugly: Attitudes Towards Invective Poetry* (Hijā') *in Classical Arabic Literature* (Leiden: Brill, 1988).

——— 'The Conceit of Pen and Sword: On an Arabic Literary Debate', *Journal of Semitic Studies* 32 (1987): 329–61.

——— *Of Dishes and Discourse: Classical Arabic Literary Representations of Food* (Richmond: Curzon, 2000).

——— 'Edible Fathers and Mothers: Arabic *Kunyas* used for Food', in *El banquete de las palabras: la alimentación de los textos árabes*, ed. Manuela Marín and Cristina de la Puente (Madrid, Spain: Consejo Superior de Investigaciones Científicas, 2005): 105–20.

——— 'Fools and Rogues in Discourse and Disguise: Two Studies', in *Sensibilities of the Islamic Mediterranean*, ed. Robin Ostle (London: I. B. Tauris, 2008).

——— 'Kufa vs. Basra: The Literary Debate', *Asiatischen Studien/Etudes Asiatiques* 50 (1996): 339–62.

——— 'Mixtures of Jest and Earnest in Classical Arabic Literature', *Journal of Arabic Literature* 23/2 (1992): 83–108.

Von Grunebaum, Gustave E. 'Aspects of Arabic Urban Literature', *al-Andalus* 20 (1955): 259–81.

——— *Medieval Islam*, 2nd edn (Chicago: University of Chicago Press, 1953).

Vergados, Athanassios, *The Homeric Hymn to Hermes: Introduction, Text, and Commentary* (Berlin: Walter de Gruyter, 2012).

Waardenburg, J. D. J., 'Mustashriḳūn', in *Encyclopaedia of Islam*, 2nd edn (Brill Online, 2012).

Walzer, Richard, 'al-Fārābī's Theory of Prophecy and Divination', in *Greek into Arabic* (Cambridge, MA: Harvard University Press, 1962): 142–8.

Welsford, Enid, *The Fool: His Social and Literary History* (London: Faber and Faber, 1935).

Wensinck, A. J., 'Kunyah', in *Encyclopaedia of Islam* V, 2nd edn (Leiden: Brill, 1986): 395–6.

Wieber, Reinhard, *Das Schachspiel in der arabischen Literatur von den Anfängen bis zur zweiten Hälfte des 16. Jahrhunderts* (Walldorf-Hessen: Verlag für Orientkunde, 1972).

Willeford, William, *The Fool and his Scepter: A Study in Clowns and Jesters and Their Audience* (Evanston: Northwestern University Press, 1969).

Xenophon, *Symposium*, in O. J. Todd (trans.), *Xenophon in Seven Volumes* IV (Cambridge, MA: Harvard University Press, 1979).

Yāqūt ibn ʿAbd Allāh al-Ḥamawī, *Muʿjam al-buldān* I, II, IV (Leipzig: F. A. Brockhaus, 1866).

——— *Muʿjam al-udabāʾ* (Beirut: Dār al-Gharb al-Islāmī, 1993).

Yusuf Ali, Abdullah (trans.), *The Holy Quran* (Beirut: Dār al-ʿArabiyya, 1968).

Al-Zubayr ibn Bakkār, *al-Akhbār al-muwaffaqiyyāt* (Qum: Manshūrāt al-Sharīf al-Raḍī, 1995).

Index